D0909064

DATE DUE

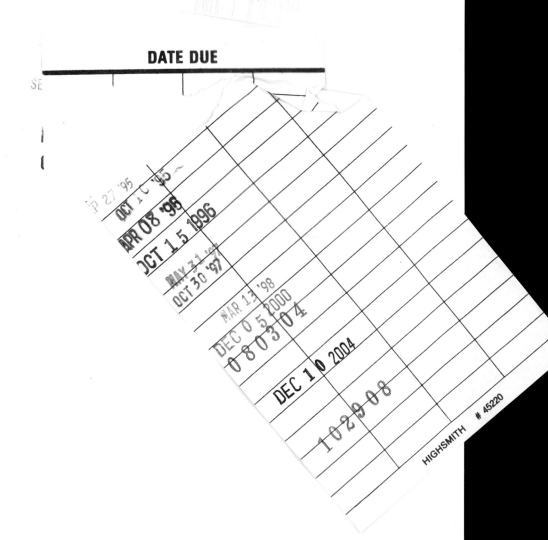

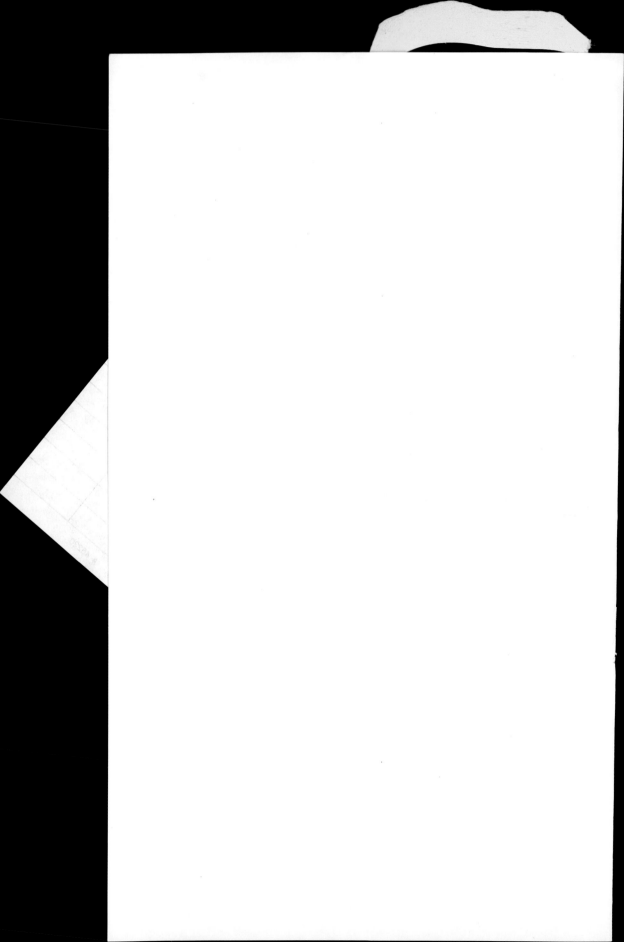

Integrated
STORMWATER
MANAGEMENT

Edited by

Richard Field
Marie L. O'Shea
Kee Kean Chin

LEWIS PUBLISHERS
Boca Raton   Ann Arbor   London   Tokyo

Library of Congress Cataloging-in-Publication Data

Integrated stormwater management / edited by
  Richard Field, Marie L. O'Shea, Kee Kean Chin.
    p. cm.
  Includes bibliographical references and index.
  ISBN 0-87371-805-4
    1. Storm sewers.   2. Water quality management.   3. Urban runoff.
I. Field, Richard.   II. O'Shea, Marie L.   III. Chin, K. K. (Kee Kean)
TD665.I54   1993
628′.21—dc20   92-533
                                                                          CIP

Direct all inquiries to CRC Press, Inc., 2000 Corporate Blvd., N.W., Boca Raton, Florida 33431.

PRINTED IN THE UNITED STATES OF AMERICA
1 2 3 4 5 6 7 8 9 0
Printed on acid-free paper

EDITORS

Richard Field received his Master's degree in Environmental Engineering from New York University in 1963. Mr. Field has been working in Environmental Engineering for 29 years, is a registered Professional Engineer in the states of New York and New Jersey and a member of Chi-Epsilon, the National Civil Engineering Honor Fraternity; and International Liaison in Urban Storm Drainage, International Association on Water Pollution Research.

He has been in the Environmental Protection Agency's (EPA) National Urban Storm and Combined Sewer Overflow (CSO) Management and Pollution Control Technology Research, Development, and Demonstration Program located at the Risk Reduction Engineering Laboratory's Edison, NJ, field station since June 1970 and is Chief of Program. Prior to his appointment with the EPA, Mr. Field served as Special Assistant for Environmental Engineering responsible for all environmental pollution problems in the First and Third Naval Districts; Chief of the Interceptor Maintenance and Plant Services Section; New York City Department of Environmental Protection; and Staff Engineer with Nussbaumer, Clarke and Velzy, Sanitary Engineering Consultants.

Mr. Field has received numerous outstanding achievement awards and citations for on-the-job performance and technological contribution including the EPA Bronze Medal, the ASCE State-of-the-Art Civil Engineering Award, two New York Water Pollution Control Association's Awards for excellence in technological advancement, and a first level EPA Scientific and Technological Achievement Award. He has written , presented, and/or published more than 425 papers/reports/books, some of which are internationally recognized publications in his field. Mr. Field is a national expert on environmental damages from and pollution control of salt (sodium chloride) and a host of stormwater and CSO best management and pollution control practices, including authorship and project officership of various national publications/projects.

Mr. Field is listed in *Who's Who in Engineering* and *Who's Who in Technology Today*.

Marie L. O'Shea received her Ph.D. in Physical Chemistry from New York University in 1990 and her B.A. in Chemistry from Reed College. She also was a Postdoctoral Fellow in the U.S. EPA Office of Research and Development's Storm and Combined Sewer Control Program from 1990 to 1992. Her current position is as Project Scientist for the New York City Department of Environmental Protection's Marine Sciences Section in the Bureau of Clean Water.

Dr. O'Shea's professional interests and work with the City of New York include investigating specific water quality problems in the NYC area including pathogens in stormwater, and dissolved oxygen depletion and PAH contamination in the New York Harbor Estuary.

Dr. O'Shea is a member of Phi Beta Kappa, American Chemical Society, New York Academy of Sciences, and Water Pollution Control Federation.

Kee Kean Chin received his Ph.D. from New York University. He is Professor of Civil Engineering at the National University of Singapore where he teaches courses in environmental engineering. He has published numerous papers and reports. As a registered Professional Engineer in Singapore and Malaysia, Dr. Chin has served as consultant to governmental bodies and private consulting firms on specialized civil and environmental engineering problems.

PREFACE

It is an honor for us to have this opportunity to present this book on Integrated Stormwater Management.

When the U.S. Environmental Protection Agency began its research work in urban storm-generated pollution control and stormwater management in 1965, the field was in its infancy. It is gratifying to see this field growing and gaining the international recognition it deserves.

Abatement or prevention of pollution from storm-generated flow is one of the most challenging areas in the environmental engineering field. The facts of life — from an engineering standpoint — are difficult to face in terms of design and cost. Operational problems can be just as foreboding.

The full impacts of "marginal" pollution, particularly that caused by uncontrolled overflows, must be recognized now and planning initiated to improve sewage system efficiencies and bring all wastewater flows under control. Municipal programs with this objective cannot begin too soon because corrective action is time-consuming. Efforts devoted to improved sewerage systems will pay significant dividends in complete control of metropolitan wastewater problems and pollution abatement. Research and development are making available important answers on the most efficient and least costly methods needed to restore and maintain water resources for maximum usefulness to man.

It is clear that abatement requirements for storm-flow pollution are forthcoming. Already, federal and local governments have promulgated wet-weather flow treatment and control standards. Now developed and developing regions can take a crucial opportunity and assess what has transpired around the globe, and determine their own best water management strategy.

To exemplify this, one can consider the water pollution control efforts in the United States. Historically, that nation has always approached water pollution control in a series and segmented manner with respect to time and pollutant sources, respectively. The result is that they are still fighting the problem after more than 60 years of effort and billions of dollars of expenditures. Initially, they abated sanitary sewage; first with primary treatment, and later only after a long time, with secondary treatment. Somewhere in between attempts to control sanitary sewage, industrial wastewater control became a requirement; however, pretreated industrial wastewaters are still released during an overflow event. Only recently were we forced to control combined sewer overflow (CSO), and now we are faced with requirements to abate separate-stormwater pollution. The aforementioned historical approach to water pollution control has taken a very long time, and only after trial and error of each individualized and fragmented approach was it learned that receiving water pollution problems remain. If instead, an entire watershed or multi drainage area analysis was conducted earlier, a determination could have been made of the overall pollution problem in the receiving-water bodies, the pollution sources (or culprits) contributing to the problem, and an optimized, integrated, area-wide program to correct the problem.

After the macro- (or large-scale watershed) analysis is conducted, an optimized determination of what sources to be abated (or where to spend the monies) will be made. Then, with the resulting information, a micro- (or drainage area/ pollutant source and control) analysis can be performed.

Toxicants should also be part of the study. Past research has shown that storm-flow toxicants and resultant toxicity can significantly affect health and the environment. Regulations for toxicants control have been promulgated and will become more demanding. The control system designs should at least be made flexible to treat toxicants once toxicants control requirements are enforced.

There is one other important consideration that must be made, i.e., the reuse and reclamation of stormwater for such beneficial purposes as aesthetic and recreational ponds, groundwater recharge, irrigation, fire protection, and industrial water supply.

An optimal approach to integrated stormwater management is a total watershed or basin-wide analyses including a macro- or large-basin-scale evaluation interfaced with a descretized micro- or small-catchment-scale evaluation involving the integration of: (1) all catchments or drainage areas, tributaries, surrounding water bodies, and groundwater; (2) all pollutant source areas, land uses, and flows, i.e., combined-sewer drainage areas, separate-storm drainage areas, including their dry-weather discharges containing unauthorized or inappropriate cross-connections, existing water pollution control plant effluents, industrial-wastewater discharges, discharges from other land uses, and air pollution fall out; and (3) added storm-flow sludge and residual solids handling and disposal. Flood and erosion control along with reuse and reclamation technology must also be integrated with pollution control, so that the retention and drainage facilities required for flood and erosion control can be simultaneously designed or retrofitted for pollution control and stormwater reclamation.

In conclusion, knowledge of interconnecting basinwide waters and pollutant loads affecting the receiving-water body and the subsurface and groundwater will result in knowing how to get the optimum water resource and pollution abatement and a much more expedient and cost effective water management programs.

Richard Field
Kee Kean Chin
Marie L. O'Shea

CONTRIBUTORS

M. Adams
Staff Engineer
Cahill Associates
West Chester, PA

Ali Ayyoubi
Department of Civil Engineering
University of Alabama
Birmingham, AL

Ronny Berndtsson
Department of Water Resources
 Engineering
University of Lund
Lund, Sweden

M. Rose Byrne
Department of Civil Engineering
University of Virginia
Charlottesville, VA

T. H. Cahill
President
Cahill Associates
West Chester, PA

Kee Kean Chin
Department of Civil Engineering
National University of Singapore
Kent Ridge, Singapore

E. L. Combes
Principal Engineer
Acer Consultants, Ltd.
Guildford, U.K.

Elvidio V. Diniz
President
Research Technology, Inc.
Albuquerque, NM

Magnus Enell
Swedish Environmental Research
 Institute
Stockholm, Sweden

Chi-Yuan Fan
Risk Reduction Laboratory
U.S. EPA
Edison, NJ

Richard Field
Storm and Combined Sewer
 Research Program
U.S. EPA
Edison, NJ

Jonathan A. French
Camp Dresser & McKee Interna-
 tional, Inc.
Cambridge, MA

Wolfgang F. Geiger
Professor
Department of Civil Engineering
Universitat Gesamthochule Essen
Essen, Germany

S. Halsey
Staff Scientist
Division of Coastal Resources
Office of Regulatory Policy
New Jersey Department of Envi-
ronmental Protection and Energy
Trenton, NJ

Brendon M. Harley
Camp Dresser & McKee Interna-
tional, Inc.
Cambridge, MA

Moh Wung Hee
Chief Engineer
Drainage Department
Ministry of the Environment
Singapore

William Hogland
Department of Water Resources
Engineering
University of Lund
Lund, Sweden

Atsushi Ichiki
Department of Civil Engineering
Ritsumeikan University
Kitaku, Kyoto, Japan

Michael A. Kasnick
Department of Civil Engineering
University of Virginia
Charlottesville, VA

Melinda Lalor
Department of Civil Engineering
University of Alabama
Birmingham, AL

P. T. Lam
Water Department
Public Utilities Board
Singapore

Magnus Larson
Department of Water Resources
Engineering
University of Lund
Lund, Sweden

C. H. Lim
Water Department
Public Utilities Board
Singapore

J. McGuire
Staff Engineer
Cahill Associates
West Chester, PA

Hiroyuki Miura
Department of Civil Engineering
University of Kansai
Suita, Osaka, Japan

Wan M. Nawang
Chairman
Department of Hydraulics and
Hydrology
Faculty of Civil Engineering
University of Teknologi Malaysia
Johor, Malaysia

A. Neysadurai
Camp Dresser & McKee Interna-
tional, Inc.
Singapore

Janus Niemczynowicz
Department of Water Resources
Engineering
University of Lund
Lund, Sweden

S. L. Ong
Department of Civil Engineering
National University of Singapore
Kent Ridge, Singapore

Marie O'Shea
New York City Department of
  Environmental Protection
Bureau of Clean Water
Marine Sciences Section
Wards Island, NY

Mana Patarapanich
Senior Lecturer
University of Technology
Sydney, Australia

Robert Pitt
Department of Civil Engineering
University of Alabama
Birmingham, AL

Euring M. J. Slipper
Water Research Centre
Wiltshire, U.K.

R. P. M. Smisson
Director
HIL Technology, Inc.
Scarborough, ME

C. Smith
Staff Engineer
Cahill Associates
West Chester, PA

Yashuhiko Wada
Professor
Department of Civil Engineering
University of Kansai
Suita, Osaka, Japan

Nancy J. Wheatley
Unit Chief
Wastewater Engineering
Massachusetts Water Resources
  Authority
Boston, MA

S. Whitney
Assistant Director
Division of Coastal Resources
Office of Regulatory Policy
New Jersey Department of Envi-
  ronmental Protection and Energy
Trenton, NJ

Michael Wong
Investigations Engineer
Hornsby Shire Council
New South Wales, Australia

C. H. Woolhouse
Department of Civil Engineering
Ritsumeikan University
Kita-ku, Kyoto, Japan

Kiyoski Yamada
Professor
Department of Civil Engineering
Ritsumeikan University
Kita-ku, Kyoto, Japan

Masaharu Yoshitomi
Department of Civil Engineering
Ritsumeikan University
Kita-ku, Kyoto, Japan

Shaw L. Yu
Department of Civil Engineering
University of Virginia
Charlottesville, VA

CONTENTS

CATCHMENT PLANNING
AND MANAGEMENT

1 STORM AND COMBINED SEWER OVERFLOW: AN OVERVIEW OF EPA'S RESEARCH PROGRAM

1.1 INTRODUCTION

The Storm and Combined Sewer Pollution Control Research, Development, and Demonstration Program (SCSP) was initiated back in 1964. Congress acknowledged the problem 26 years ago by authorizing funds under the Water Quality Act of 1965 for researching ways of stormwater pollution management. The research effort is directed by the Storm and Combined Sewer Technology Program located in Edison, New Jersey. About 300 projects totaling approximately $150 million have been awarded under the U.S. Environmental Protection Agency (EPA) research program which resulted in approximately 320 final reports. More than 100 conference papers and over 100 articles and in-house reports have been presented and published, respectively, by the Program. The goal has been user assistance with emphasis on planning and design oriented material.

The mission of the SCSP was to develop methods for controlling pollution from urban stormwater discharges and combined sewer overflows (CSO), and excessive inflow and infiltration (I/I).

The program had two facets. The first was problem definition that led to the second — development of effective control alternatives.

The program has been involved in the development of a diverse technology including pollution-problem assessments/solution methodology and associated instrumentation and stormwater management models, best management practices (BMP), erosion control, infiltration/inflow (I/I) control, CSO and stormwater control-treatment technology and associated sludge and solids residue handling and disposal methods, and many others. This report covers SCSP products and accomplishments in these areas covering 18 years of efforts. The vastness of the

3

program makes it difficult to allow complete coverage. Therefore program outputs and developments will be selectively emphasized.

1.2 POLLUTION PROBLEM ASSESSMENT

1.2.1 Background

The background of sewer construction led to the present urban runoff problem. Early drainage plans made no provisions for storm-flow pollutional impacts. Untreated overflows occur from storm events giving rise to the storm-flow pollution problem.

It was recently estimated it would cost the U.S. approximately $100 billion and $200 billion for CSO and urban stormwater control, respectively.

Simply stated the problem is: when a city takes a shower what do you do with the dirty water?

Three types of discharges are involved: (1) combined sewer overflow (CSO), which is a mixture of storm drainage and municipal wastewater, and which also includes dry-weather flow (DWF) discharged from a combined sewer due to clogged interceptors, inadequate interceptor capacity, or malfunctioning regulators; (2) storm drainage from separate storm systems either sewered or unsewered; and (3) another form of CSO, overflow from sanitary lines infiltrated with stormwater.

1.2.2 Characterization

The problem constituents in overflows are visible matter, infectious (pathogenic) bacteria and viruses, oxygen demanding matter and solids, and in addition, include nutrients, and toxicants (e.g., heavy metals, pesticides, and petroleum hydrocarbons) .

The average 5-day biochemical oxygen demand (BOD) concentration in CSO is approximately one half the raw sanitary sewage BOD, but storm discharges must be considered in terms of their shock loading effect due to their relative magnitude. Urban runoff flow rates from an average storm intensity of 0.1 in./h are five to ten times greater than the DWF from the same area. Likewise a not uncommon rainfall intensity of 1.0 in./h will produce flowrates 50 to 100 times DWF. Even separate storm wastewaters are significant sources of pollution, typically characterized as having solids concentrations equal to or greater than those of untreated sanitary wastewater and BOD concentrations approximately equal to those of secondary effluent. The bacterial and viral pollution problem from wet-weather flow (WWF) is also severe.

The quality and quantity characterization of WWF is necessary for problem assessment, planning, and design. Summaries of characterization data from many research studies are available.[1-3] The average pollutant concentrations for urban runoff and CSO are compared to background pollution and sanitary sewage in Table 1.[2]

Since 1974, the program supported the urban rainfall-runoff-quality database for two important data requirements: characterization, and calibration and verification of models.[4] This project was initiated to bring together the many widely scattered data sources.

1.2.3 Case Studies

A few municipal studies can serve to exemplify the problem. In Northampton, England, it was found that the total mass of BOD emitted from CSO over a two-year period was approximately equal to the mass of BOD emitted from the secondary sewage treatment plant effluent. And that the mass emission of suspended solids (SS) in CSO was three times that of the secondary effluent. In Buffalo, New York, a study concluded that 20 to 30% of the DWF solids settled in the combined sewer that was subsequently flushed and bypassed during high-velocity storm flows.

A study in Durham, North Carolina, has shown that after providing secondary treatment of municipal wastes, the largest single source of pollution from the 1.67-mile watershed is separate urban runoff without the sanitary constituent. When compared to the raw municipal waste generated within the study area the annual urban runoff of chemical oxygen demand (COD) was equal to 91% the raw sewage yield; the BOD yield was equal to 67%; and the SS yield was 20 times that contained in the raw municipal wastes.

From an inhouse project, preliminary screening of urban wet-weather discharges from 24 samples from 9 urban areas found approximately one half of the 129 priority pollutants. The heavy metals were consistently found in all samples. Polynuclear aromatic hydrocarbons (PAHs), from petroleum, were the most frequently detected organics followed (in order) by phthalate esters, aromatic hydrocarbons, halogenated hydrocarbons, and phenols. A few other EPA studies also indicated that CSO and stormwater contain significant quantities of priority pollutants.

A project in Syracuse, New York, used the Ames test to evaluate urban runoff and CSO mutagenicity.[5] Detectable responses have been obtained on 22% of the samples. It is significant that some mutagenic substances are present with a potential for entering the food chain.

Additional investigation of the significance of toxic pollutants with regard to their health effects and ecosystem effects is being conducted along with an evaluation of the removal capacity of alternative treatment technologies for these toxicants and comparison of their effectiveness with estimated removal needs to meet water quality goals. From this comparison, treatment and control for toxic substances removal will need to be developed.

Indicators, such as fecal coliform, have long been known to be present in stormwater discharges in densities sufficient to cause contravention of standards. A study in Baltimore, Maryland, identified actual pathogens and enteroviruses in storm sewer discharges.[6] Cross-connections from sanitary sewers were strongly implicated as the major cause. Obviously, this problem is not

Table 1. Comparison of Typical Values for Storm Flow Discharges[a]

	TSS	VSS	BOD	COD	Kjeldahl Nitrogen	Total Nitrogen	PO_4-P	OPO_4-P	Lead	Fecal Coliforms
Background levels	5–100	—	0.5–3	20	—	0.05–0.5[b]	0.01–0.2[c]	—	<0.1	—
Stormwater runoff	415	90	20	115	1.4	3–10	0.6	0.4	0.35	14,500
Combined sewer overflow	370	140	115	375	3.8	9–10	1.9	1.0	0.37	670,000
Sanitary sewage	200	150	375	500	40	40	10	7	—	—

[a] All values mg/L except fecal coliforms which are organisms/100 mL.
[b] NO_3 as N.
[c] Total phosphorus as P.

isolated to Baltimore. For instance, two surveys in Canada found that 13% and 5% of the houses had illicit sanitary connections to separate storm sewers, respectively. At this juncture, because of the high expenses involved for disinfection it is important to mention that better indicator organisms of human disease potential are needed since the conventional indicators, e.g., coliform can come from animal fecal matter and soil in the runoff whereas in sanitary flow it is principally from human enteric origin. Perhaps direct pathogen measurement is best.

1.2.4 Receiving-Water Impacts

Knowledge of the receiving-water impacts resulting from urban wet-weather discharges is a basis for determining the severity of problems and for justifying control. Program studies of receiving-water impacts are described in a proceedings from a national conference and in a journal paper.[7,8]

Oxygen Demand Loads

Under certain conditions, storm runoff can govern the quality of receiving waters regardless of the level of DWF treatment provided. Based on national annual mass balance determinations, (Table 2) wet-weather oxygen demand loads are greater than the dry-weather (sanitary sewer) loads from the same areas and ten times greater during storm-flow periods.[9,10] Hence, control of storm runoff pollution is a viable alternative for maintaining receiving-water quality standards.

Aesthetic Deterioration and Solids

Stormwater conveys debris and solids to receiving waterbodies. This material can either disperse, float, or wash ashore onto beaches or embankments, or eventually settle, creating such nuisances as odors and toxic/corrosive atmospheres from bottom mud deposits, and aesthetic upsets either in general appearance (dirty, turbid, cloudy) or in the actual presence of specific objectionable items (floating debris, oil films, sanitary discards/fecal matter, scum or slimes, tires, timber, etc.).

Coliform Bacteria and Pathogenic Microorganisms

Excess concentrations of bacterial indicator organisms in urban runoff will hinder water supply, recreational, and fishing/shellfishing use of the receiving water.[6-8] Elevated coliform levels in Mamaroneck Harbor, New York and subsequent beach closings have been linked to stormwater runoff. Stormwater discharges from the City of Myrtle Beach, South Carolina, directly onto the beach showed high bacterial counts for short durations immediately after storm

Table 2. National Annual Urban Wet-/Dry-Weather Flow (WWF/DWF) BOD$_5$ and COD Comparisons [9,10]

Type	Percent of Developed Area	Annual DWF		Annual WWF		Percent WWF	
		BOD$_5$ x 10^6 lb	COD x 10^6 lb	BOD$_5$ x 10^6 lb	COD x 10^6 lb	BOD$_5$	COD
Combined sewer	14.3	340	910	880	2640	72	74
Storm sewer	38.3	710	1890	440	2500	36	57
Unsewered	47.4	310	830	360	2250	54	73
Totals	100	1360	3630	1680	7390	55	67

events. In many instances these counts violated EPA recommended water quality criteria for aquatic life and contact recreation. In Long Island, New York, stormwater runoff was identified as the major source of bacterial loading to marine waters and the indirect cause of the closing of about one-fourth of the shellfishing area.

Biological Impacts

An investigation of aquatic and benthic organisms in Coyote Creek, San Jose, California, found a diverse population of fish and benthic macroinvertebrates in the nonurbanized section of the creek as compared to the urbanized portion, which was completely dominated by pollution tolerant algae, mosquito fish, and tubificid worms.[7,8,11] In the State of Washington similar results were found in a Lake Washington project where bottom organisms (aquatic earthworms) near storm sewer outfalls were more pollution-tolerant relative to those at a distance from these outfalls.[7,8,12] Aquatic earthworm numbers and biomass were found to be enhanced within the zone of influence of the monitored storm drain in the lake.

Toxicity

Toxicity problems can result from minute discharges of metals, pesticides, and persistent organics which may exhibit subtle long-term effects on the environment by gradually accumulating in sensitive areas. A large data-base exists that identifies urban runoff as a significant source of toxic pollutants, e.g., New York Harbor receives metals from treatment plant effluents, combined sewer overflows, separate storm sewer discharges, and untreated wastewater.[7,8] As seen in Table 3, urban runoff is the major contributor of heavy metals to the harbor. Table 4 shows the total annual mass of selected constituents from a storm overflow point in Seattle, Washington.[7,8,12]

A high percentage of the heavy metals and toxic materials is associated with the SS or particulates which tend to concentrate in the sediment. This association is beneficial in terms of control and treatment since it is easier to separate pollutants attached to SS.

Sediment samples were analyzed for metals, organic carbon, phosphorous, chlorinated hydrocarbons, and polychlorinated biphenyls (PCBs). As can be seen from Figure 1, a composite index to assess wet-weather impacts was 16 times the minimum background control value. Also, pesticide levels in sediments along the Seattle shoreline of Lake Washington were up to 37 times background concentrations.

In the previously mentioned San Jose project,[11] urban sediment compared to nonurban sediment from Coyote Creek contained higher peak concentrations with up to ten times more lead — 400 mg/kg vs. 40 mg/kg, nine times more arsenic — 13 mg/kg vs. 1.5 mg/kg, up to four times more BOD — 2,900 mg/kg

Table 3. Metals Discharged in Harbor from New York City Sources[7,8]

Source	Copper	Chromium	Nickel	Zinc	Cadmium
Plant effluents	1,410	780	930	2,520	95
Runoff[a,b]	1,990	690	650	6,920	110
Untreated wastewater	980	570	430	1,500	60
Total weight, lb/day	4,380	2,050	2,010	10,940	265
Average concentration, mg/L	0.25	0.12	0.11	0.62	0.015

[a]   In reality, shock-load discharges are much greater.
[b]   Runoff data includes separate storm sewer drainage and wet-weather combined sewer overflows (CSO).

Table 4. Total Vs. Particulate Mass from Storm Sewer Overflow Point; Lake Washington, Seattle, Washington[7]

	Selected Storm Drain Point	
Variable	Total Mass, in Pounds	Particulate Mass, in Pounds
Suspended solids	4924	4924
Copper	2.55	1.64
Lead	13.29	11.7
Zinc	6.03	3.87
Aluminum	213.8	207
Organic carbon	658	370
Total phosphorous	19.2	8.93
Oils and greases	249	not applicable
Chlorinated hydrocarbons	not determined	0.854 G

vs. 925 mg/kg, and four times more ortho phosphates — 6.7 mg/kg vs. 1.8 mg/kg. Lead concentrations in urban samples of algae, crawfish, and cattails were two to three times greater than in nonurban samples, while zinc concentrations were about three times the nonurban concentrations. Bioaccumulation of lead and zinc in the organisms compared to water column concentrations was at least 100 to 500 times greater.

Petroleum hydrocarbons, particularly the polynuclear aromatics, are suspected carcinogens. At New York City's Newton Creek treatment plant, 24,000 gal of oil and grease, equivalent to a moderate spill, were bypassed during one 4-h storm.[7] A study of Jamaica Bay, New York, found that 50% of the hexane extractable material contributed to the bay is due to wet-weather

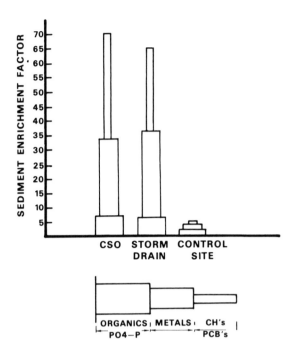

Figure 1. Urban sediment enrichment. Lake Washington, Seattle, Washington.

overflows.[7] The major source of petroleum contamination in Jamaica Bay was shown to be waste crankcase oil.[13] This is in agreement with studies of Delaware Bay.[13] Petroleum hydrocarbons and associated aromatic hydrocarbons are a cause of ecosystem degradation in New York Bight.[13] Accumulation of polynuclear aromatics in sediments eventually may prove harmful to benthic communities in the Bight. It is clear from Table 5 that urban runoff is the major factor to be considered for control of hydrocarbons.

Over 86% of the total hydrocarbons in Philadelphia, Pennsylvania, storm runoff was associated with particulates, a distribution that probably is typical of most urban areas.[13] Therefore, instream solids separation being designed and considered for separate stormwater systems will result in substantial lowering of nonpoint petroleum hydrocarbon inputs, provided the solids are disposed of elsewhere.

Sediment

Direct evidence has been obtained (from the Milwaukee River project)[14] of how a disturbed benthos depletes dissolved oxygen (DO) from the overlying waters. Previously mentioned (and other) studies have also shown that stormwater

**Table 5. Estimated Sources of Petroleum
Hydrocarbons In Delaware Bay[13]**

	Without Efficient Controls (lb/day)	With Efficient Controls (lb/day)
Spills	6,000	6,000
Municipal	7,700–15,700	2,000
Refineries	24,300	2,000
Other industrial	8,800	6,200
Urban runoff	10,600	10,600

discharges and CSO's adversely affect sediment by toxics' enrichment and resultant biological upsets.[7,8,11,12] Since particulate matter in untreated stormwater discharges and CSOs is larger, heavier, and in significant quantities when compared to treated sanitary effluent, more needs to be known about the size and inertial characteristics, fate, and transport of settlable and separable materials. Hydrodynamic solids separation and sediment transport routines must be added to receiving water models to take care of the neglected or presently omitted significant particulate- and bed-flow fields.

1.3 SOLUTION METHODOLOGY

The concept of a simplified continuous receiving water quality model was developed in the nationwide evaluation of CSO's and urban stormwater discharges project and refined into a user's manual during a subsequent project.[15,16] This model, termed "Level III — Receiving", permits preliminary planning and screening of area-wide wastewater treatment alternatives in terms of frequency of water quality violations based on time and distance varying DO profiles. Figure 2 represents a hypothetical example of the type of analysis facilitated by this model. This case is for DO; actual studies should include other parameters and should represent at least one year of continuous data.

Using this analysis a truer cost-effectiveness comparison can be made based on the duration of the impact and associated abatement costs, e.g., if a 5 mg/L DO is desired in the receiving water 75% of the time, an advanced form of wet-weather treatment or primary wet-weather treatment integrated with land management is required. The latter is the most cost effective at $3,000,000. This or similar tools will aid in setting cost-effective standards as well as the selection of alternatives.

Also, a general methodology has been developed for evaluating the impact of CSOs on receiving water and for determining the abatement costs for various water quality goals.[17] It was developed from actual municipal pollution control

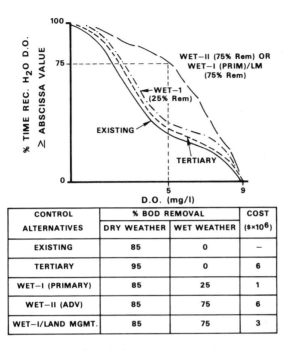

Figure 2.   Hypothetical example solution methodology.

facility planning experience in Onondaga Lake in Syracuse, New York. An important goal of studies to determine the impact of waste discharges on a receiving water is to predict the waste loads that can be assimilated without violation of water quality standards so that a loading curve such as shown in Figure 3 can be defined.

This figure shows the potential effect storm loads may have in violating a 5 mg/L DO standard after dry-weather treatment is upgraded. It further implies that CSO pollution loads should be abated next, since they are the easiest of the storm loads to control and capture; and in this case would reduce loads to meet the water quality goal.

In addition, a methodology for defining criteria for wet-weather quality standards has been developed.[18,19] In recognition of an important gap in the developed methodologies, the duration of water quality standards vs. species survival was taken into consideration.

1.4 USER'S ASSISTANCE TOOLS

User's assistance tools include instrumentation, stormwater management models, manuals of practice (MOP), methodologies, compendiums, and state-of-the-art (SOTA) reports.

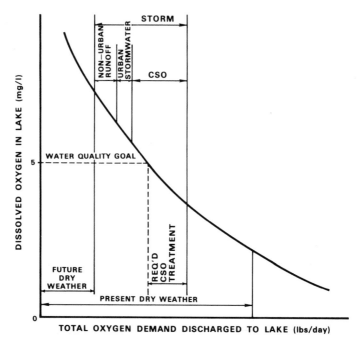

Figure 3.   Typical loading curve relating pollutant load to water quality response.

1.4.1 Instrumentation

Storm-flow measurement is essential for process planning, design, control, evaluation, and enforcement. Sampling devices do not provide representative aliquots. Conventional flowmeters apply to steady-state flows and not to the highly varying storm flows.

Flowmeters have been developed to overcome these adverse storm conditions.[20,21] A prototype sampler for capturing representative solids in storm flow has also been developed and a design manual is available.[22] This gave manufacturers the incentive to perfect samplers by increasing intake velocities and other ways. SOTA reports are available for flow measurement and sampling.[23,24] Because storm-flow conditions are extremely adverse, the manuals and instruments developed are useful for monitoring all types of flow.

1.4.2 Simulation Models

The program has fostered the development of models for assessment, planning, design, and control of urban stormwater pollution. Program thinking on urban water management analysis involves four levels of evaluation, ranging from simple to complex, that can be worked together.

The various levels of the stormwater management model (SWMM) are the most significant model products in terms of past resources and overall popularity. SWMM is one of the most widely used urban models and its benefits for planning and design have been demonstrated. It has been employed by consulting engineers to design sewers and to analyze pollution control alternatives.

There have been significant enhancements of SWMM. Probably the most significant is Version III,[25,26] which includes a flexible physically-based storage and treatment routine that provides estimates of treatment (by settling) in storage basins. We have just integrated SWMM with a GIS; and have plans to incorporate toxicants characterization and associated treatability databases into SWMM.

Documentation and user's manuals are available for all SWMM levels, including three continuous stormwater planning models.[27-30] Operational models that have been implemented in Detroit, Michigan;[31] Minneapolis, Minnesota;[32] Seattle, Washington;[33] and San Francisco, California, produce control decisions during storm events.

1.4.3 Reports

A major emphasis of the Program was solution methodology through developments of SOTA reports, manuals of practice (MOPs), and user's manuals.

The SOTA texts, user's guides, and the assessment on urban stormwater technology are excellent documents.[34,35] Separate engineering manuals are available for storm flowrate determination,[36,37] porous pavement design,[38-40] cost estimating,[10,41,42] storm sewer design,[43-45] planning and design guidance,[46] and for conducting stormwater studies.[47] Seminar proceedings with themes of "modeling, design, operation, and costs" have been published. The SOTA document on particle size and settling velocity offers significant information for solids treatability and their settlement in receiving waters, important areas overlooked in planning and design.[48] An excellent film is being distributed by the General Services Administration (GSA) National Audio Visual Center which covers the EPA Storm and CSO Research Program, and in particular full-scale control technologies.[49]

A report entitled, "Urban Stormwater Management and Technology: Case Histories",[50] presents 12 case histories that represent the most promising approaches to CSO and stormwater control. The case histories were developed by evaluating operational facilities that have significant information for future guidance.

Three illustrative methodologies for conducting CSO facility planning have been published having significant guidance.[51-53]

1.5 MANAGEMENT ALTERNATIVES

The next major Program area is management alternatives. First is the choice of where to attack the problem; at the source by land management, in the collection system, or off-line by storage. We can remove pollutants by treatment and by employing integrated systems combining control and treatment.

1.5.1 Land Management

Land management includes structural, semistructural, and nonstructural measures for reducing urban and construction site stormwater runoff and pollutants before they enter the downstream drainage system. The following various concepts have been fostered by the SCSP.

Land Use Planning

Traditional urbanization upsets the natural hydrologic and ecological balance of a watershed. The degree of upset depends on the mix, location, and distribution of the proposed land use activities. As man urbanizes, the receiving waters are degraded by runoff from his activities. The goal of urban development resources planning is a macroscopic management concept to prevent problems from shortsighted planning. New variables of land usage and its perviousness, population density, and total runoff control must be considered and integrated with desired water quality by land-use planners.

Natural Drainage

Natural drainage will reduce drainage costs and pollution, and enhance aesthetics, groundwater supplies, and flood protection. A project near Houston, Texas, focused on how a "natural-drainage system" integrates into a reuse scheme for recreation and aesthetics.[50] Runoff flows through vegetative swales and into a network of wet-weather ponds, strategically located in areas of porous soils. This system retards the flow of water downstream, preventing floods by development, and enhances pollution abatement.

An interesting technological answer to the problem of preserving pervious areas is using an open-graded asphaltic-concrete as a paving material. This will be discussed later under the subsection "Porous Pavements".

Multipurpose Detention/Retention

Multipurpose detention/retention and drainage facilities, and other management techniques required for flood and erosion control, can be simultaneously designed or retrofitted for pollution control. Retention on-site or upstream can provide for the multibenefits of aesthetics, recreation, recharge, irrigation, or other uses. An existing detention basin can be retrofitted to enhance pollution control by limiting or eliminating the bottom effluent orifice and by routing most or all of the effluent stormwaters through a surface overflow device, e.g., a weir or standpipe drain. This will induce solid-liquid separation by settling and enable entrapped solids and floatables to be disposed of at a later time without causing downstream receiving water pollution.

Major-Minor Flooding

By utilizing the less densely populated and less commercialized upstream, upland drainage areas for rainwater impoundment for the more intense storms, the relatively (and significantly) more costly and upsetting downtown downstream flooding can be eliminated or alleviated. The multibenefits of pollution abatement and a reduced need for larger pipes downstream will also be gained. The "major-minor flooding" concept involves utilization of depression storage by brief flooding of curblines, right-of-ways, and lawn areas.

Controlled Stormwater Entry

A project in Cleveland, Ohio, demonstrated how controlling the rate at which stormwater stored upstream enters the sewerage system alleviates basement flooding and overflow pollution. The flow rate is regulated by a vortex internal-energy dissipator (Hydrobrake). This small device, which is located at the downstream end of a subsurface holding tank beneath the right-of-way, delivers a predesigned virtually constant discharge rate, compatible with the downstream sewerage system capacities and water quality objectives, regardless of head variations. This is accomplished without the need of moving parts or external energy sources.

Porous Pavement

Porous pavements provide storage, enhancing soil infiltration that can be used to reduce runoff and CSO. Porous asphalt-concrete pavements can be underlaid by a gravel base course with whatever storage capacity is desired (Figure 4). Results from a study in Rochester, New York, indicate that peak runoff rates were reduced as much as 83%.[51] The structural integrity of the porous pavement was not impaired by heavy-load vehicles. Clogging did result from sediment from adjacent land areas during construction; however, it was relieved from cleaning by flushing. The construction cost of a porous pavement parking lot is about equal to that of a conventional paved lot with stormwater inlets and subsurface piping.

A demonstration project in Austin, Texas, developed design criteria for porous pavements and compared porous asphalt pavement to six other conventional and experimental pavements.[39,40]

Surface Sanitation

Maintaining and cleaning urban areas can have a significant impact on the quantity of pollutants washed off by stormwater with secondary benefits of a cleaner and healthier environment.

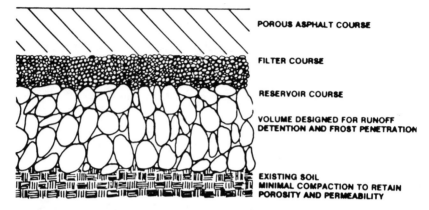

POROUS ASPHALT COURSE

FILTER COURSE

RESERVOIR COURSE

VOLUME DESIGNED FOR RUNOFF
DETENTION AND FROST PENETRATION

EXISTING SOIL
MINIMAL COMPACTION TO RETAIN
POROSITY AND PERMEABILITY

Figure 4.   Porous asphalt paving typical section.

Litter Control

Spent containers from food and drink, cigarettes, newspapers, sidewalk sweepings, lawn trimmings, and a multitude of other materials, carelessly discarded become street litter. Unless this material is prevented from reaching the street or is removed by street cleaning, it often is found in stormwater discharges. Enforcement of antilitter laws, convenient location of sidewalk waste disposal containers, and public education programs are just some of the source control measures that can be taken at the local level. While difficult to measure, the benefits include improved aesthetics and reduced pollution.

According to a recent California study,[2] litter accumulates at a rate of approximately 4 lb/person/year in urban areas. Of this total, about 1.8 lb/person/year appears between the curb lines of streets. For example, the estimated annual litter deposition for a municipality having a population of 100,000 is 400,000 lb. It was reported that about 21% of the material picked up during mechanical street sweeping was litter.

Chemical Control

One of the most often overlooked measures for reducing the pollution from stormwater runoff is the reduction in the indiscriminate use and disposal of toxic substances such as fertilizers, pesticides, oil, gasoline, and detergents.

Operations such as tree spraying, weed control, fertilization of parks and parkways by municipal agencies, and the use of pesticides and fertilizers by homeowners can be controlled by increasing public awareness of the potential hazards to receiving waters, and providing instruction as to proper use and application. In many cases over-application is the major problem, where use in moderation would achieve equal results. The use of less toxic formulations is another alternative to minimize potential pollution.

Pesticides have been detected in samples taken from several urban areas with typical loadings, including PCB, between 0.000136 to 0.012 lb/curb/mile.[2] Direct and indirect dumping and/or spills of chemicals, hazardous substances, crankcase oil, and debris into streets and gutters, catchbasins, inlets, and sewers are significant problems that may only be addressed through educational programs, ordinances, and enforcement.

Street Sweeping/Cleaning

Tests under real-world conditions in San Jose, California, showed that street cleaning can remove up to 50% of the total solids (including litter) and heavy-metal yields in urban stormwater with once or twice a day cleaning.[52] Typical street cleaning programs of once or twice a month proved ineffective. Organics and nutrients that originate primarily from surfaces upstream of streets and may be dissolved or dissolved residue, could not be effectively controlled even with intensive cleaning.

In Bellevue, Washington, conventional street cleaning proved ineffective; however, a modified regenerative air Tymco street cleaner showed promise.[53] The main purpose of street sweeping is to enhance aesthetics by cleaning up litter and coarser solids. Street cleaning is no panacea for stormwater pollution control (and is site specific dependent upon rainfall/climatic conditions), but if integrated with other methods, could reduce city-wide costs for pollution control and in general. When considering that street sweeping is used in many locations for aesthetic purposes only, it will also provide a dual benefit, i.e., low-level water pollution control, especially enhancement of receiving water aesthetics.

Deicing Practices

Effective management of street and highway deicing practices can lessen environmental and receiving water impacts, often without a substantial increase in costs. A 1973 assessment study concluded major adverse environmental effects come from sloppy salt storage and over-application, which resulted in MOPs for improvement in those areas. These manuals were recognized as highly significant. The Federal Highway Administration reprinted them and distributed approximately 10,000 copies. Recommended modifications to current deicing practices include: (1) judicious application of salt and abrasives, (2) reduced application rates (using sodium and calcium salt premixes: rates of 150 to 400 lb/lane/mile have been recommended),[2] (3) using better spreading and metering equipment and calibrating application rates, (4) prohibiting use of chemical additives, (5) providing improved (covered and/or properly drained) salt storage areas, and (6) educating the public and operators about the effects of deicing technology and the best management practices.[2] The Program work encouraged states (e.g., Wisconsin) and local governments to abate salt usage.

We found that the use of chemical additives such as cyanide, phosphate, and chromium can result in polluted snowmelt. Chromium concentrations of 3.9 mg/L have been reported.[2]

Costs associated with salting of roadways, both direct and indirect, were estimated on an annual basis for the snowbelt states.[2] A total annual cost of $3 billion was reported, of which only $200 million was associated with salt purchase and application. Other costs in the total estimate included: (1) loss and contamination of water supplies and damage to health, $150 million; (2) vegetation damage, $50 million; (3) damage to highway structures, $500 million; (4) vehicle corrosion damage, $2 billion; and (5) damage to utilities, $10 million.

1.5.2 Collection System Control

The next overall Program category, collection system controls, pertains to management alternatives for wastewater interception and transport. These include: sewer separation; improved maintenance and design of catchbasins, sewers, regulators, and tide gates; and remote flow monitoring and control. The emphasis, with the exception of sewer separation, is on optimum use of existing facilities and fully automated control.

Sewer Separation

The concept of constructing new sanitary sewers to replace existing combined sewers as a control alternative, has largely been abandoned due to enormous costs, limited abatement effectiveness, inconvenience to the public, and extended time for implementation. Again separate stormwater is a significant pollutant and sewer separation couldn't cope with this load. It is further estimated that the use of alternative measures could reduce costs to about one third of the cost of separation.

Catchbasins

In a project conducted in Boston, Massachusetts, catchbasins were shown to be potentially quite effective for solids reduction (60 to 97%).[54] Removals of associated pollutants such as COD and BOD were also significant (10 to 56% and 54 to 88%, respectively). To maintain the effectiveness of catchbasins for pollutant removal requires a municipal commitment with cleaning probably twice a year depending upon conditions. The Program developed an optimal catchbasin configuration based on hydraulic modeling.

Sewers

Manuals on new sewer design to alleviate sedimentation and resultant first-flush pollution and premature bypassing,[43] and sewer design for added CSO storage are available.[45]

Sewer Flushing

As a follow-up to an earlier study,[55] providing simple equations for predicting dry-weather deposition in sewers, a report was published showing that sewer flushing flush waves can effectively convey sewer deposits including organic matter.[56] In another study it was concluded that sewer flushing could reduce CSO control costs 7% when compared to a CSO storage/treatment and disinfection facility designed for a one-year storm.[57]

Polymers to Increase Capacity

Research has shown that polymeric injection can greatly increase flow capacity (by reducing wall friction) and thus be used to correct pollution-causing conditions such as localized flooding and excessive overflows.[58] Direct cost savings are realized by eliminating relief-sewer construction.

In-Sewer Storage and Flow Routing

Another control method is in-sewer storage and routing to maximize use of existing sewer capacity. The general approach comprises remote monitoring of rainfall, flow levels, and sometimes quality, at selected locations in the network, together with a centrally computerized console for positive regulation. This concept has proven effective in New York City, Detroit,[31] Seattle (Figure 5),[33] Cleveland, and San Francisco. Other cities are considering its use. New York City used a simple static weir that impounded upstream CSO up to a level where flooding would not be encountered. It is not a remote-controlled intelligent system; however, it provides ten million gallons of storage at practically no cost.

Although never tried, separate storm sewers and channels can also be retro-fitted with flow regulators (and sensing devices) for in-channel and in-pipe storage applying CSO in-sewer storage and routing technology and other storage facilities for ties into the existing sewage treatment system, thus making better use of facilities and lowering costs for overall water pollution control. Ties into the existing treatment system will be discussed in more detail under the subsections, "Swirl and Helical Flow Regulators/Solids Concentrators" and "Flow Regulators for Separate Stormwater Pollution Control".

Sewer System Cross-Connections

Research efforts have shown that sanitary and industrial contamination (by cross connections) of separate storm sewers is a nationwide problem. One response to this problem includes simple methods of checking for cross-connections. Investigations should be made of the drainage network, using visual observation and screening/mass balance techniques by quality sensing, to determine the sources of sanitary or industrial contamination.

Figure 5.   Computer console for augmented flow control system, Seattle, Washington.

First, stormwater outfalls can be checked by sight for discharges during DWF conditions, and if flows are noticed, the stormwater outfalls should be observed further for clarity, odor, and sanitary matter. These dry-weather discharges should then be confirmed and quantified (for relative amounts of stormwater, groundwater, sanitary, and industrial wastewater) by thermal (temperature), chemical (specific ions), and/or biochemical (BOD, COD, TOC) techniques and mass balances. Mass balances will depend on determined concentrations/values of parameters (i.e., pollutants and/or specific ions and/or temperatures) in the various potential sewer flows (i.e., stormwater, groundwater, sanitary, and industrial wastewaters). If visual outfall observations cannot be made during low tides, then upstream observations or downstream sampling (during low tide and nontidal backflow conditions) should be conducted. The drainage or sewer system flows as a branch and tree-trunk network which enables the investigators to strategically work upstream to isolate the sources of stormwater contamination or cross-connections.

Once the sources have been isolated, an analysis will have to be made to determine whether corrective action at the sources, i.e., eliminating the cross connection(s), or downstream storage/treatment (dealing with the storm sewer/channel network as though it were a combined sewer network), is most feasible. This will depend on the amount, dispersion, and size of the cross-connections.

Flow Regulators and Tide Gates

Pacesetters in the area of CSO regulator technology were the Program's SOTA and MOP.[59,60]

Conventional regulators malfunction, lack flow-control ability, and cause excessive overflows. Devices such as the fluidic regulator, and the positive control gate regulator, have been demonstrated in Philadelphia, Pennsylvania,[61] and Seattle, Washington, respectively.[33]

Swirl and Helical Flow Regulators/Solids Concentrators

The dual-functioning swirl flow regulator/solids concentrator has shown outstanding potential for simultaneous quality and quantity control.[62,63]

The swirl has been demonstrated for CSO in Syracuse, New York, and Lancaster, Pennsylvania, by the Program and elsewhere by others. The device of simple annular construction requires no moving parts (Figure 6). It controls flow by a central circular weir, while simultaneously treating combined wastewater by a "swirl" action which imparts liquid-solids separation. Tests indicate at least 50% removal for SS and BOD. Table 6 shows the Syracuse prototype results. Tankage is small compared to sedimentation making the device highly cost effective.

A helical type regulator/separator has also been developed based on principles similar to the swirl. A project in West Roxbury, Massachusetts, represents the first trial on separate stormwater.[64] There have been a number of full-scale projects throughout the country using the swirl. A complete swirl/helical design textbook has been published.[64]

Flow Regulators for Separate Stormwater Pollution Control

To protect receiving water from the effects of stormwater discharges, conventional static- or dynamic-flow regulators used for CSO control can be installed in separate storm sewers to divert stormwater to either a sanitary interceptor and/ or to a storage tank for coarse solids and floatables removal and subsequent treatment at the dry-weather plant.[63,64]

At present, there is a strong need to develop and have a reserve of control hardware for urban runoff control and to effectively reduce the associated high cost implications for conventional storage tanks, etc. It is felt that the swirl/ helical type regulators, previously applied only to CSO, can also be installed on separate storm drains before discharge and the resultant concentrate flow can be stored in relatively small tanks, since concentrate flow is only a few percent of the total flow.

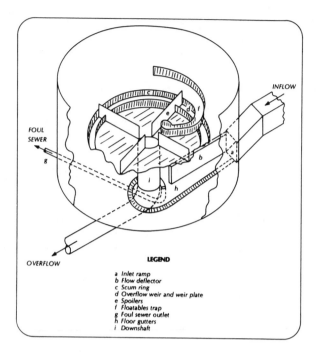

**Figure 6. Isometric view of swirl combined sewer overflow regulator/
separator.**

**Table 6. Swirl Regulator/Separator Suspended Solid
Removal: Syracuse, New York, Prototype Results**

Mass Loading (kg) per Storm			Average SS per Storm (mg/L)		
Inf.	Eff.	Removal	Inf.	Eff.	Removal
374	179	52%	535	345	36%
103	24	77%	374	165	55%
463	167	64%	342	202	41%

Stored concentrate can later be directed to the sanitary sewer for subsequent treatment during low-flow or dry-weather periods, or if capacity is available in the sanitary interceptor/treatment system, the concentrate may be diverted directly to it without storage.

This method of stormwater control (illustrated in Figure 7) is more economical than building huge holding reservoirs for untreated runoff, and offers a feasible approach to the control and treatment of separately sewered urban stormwater.[63]

Vortex Energy Dissipators

We have demonstrated vortex energy dissipators (in these cases Hydrobrakes) in Rochester, New York,[51] and Cleveland, Ohio.[65] They can be used as upstream off-line flow attenuators for controlled entry (as described previously), in-line or in-sewer flow back-up devices, and as CSO regulators. The flow rate is regulated by the vortex internal energy dissipator concept. It delivers a predesigned, virtually constant, discharge rate, compatible with the downstream sewerage system capacities and water quality objectives regardless of upstream head variations. This is accomplished without the need of moving parts, orifice closure, or external energy.

Rubber "Duck Bill" Tide Gate

Figure 8 shows a prototype rubber "duck bill" tide gate. The prevailing problems with conventional flap-type gates are their failure to close tight and the need for constant maintenance. Poor tide gate performance results in higher treatment costs, treatment plant upsets, and greater pollutional loads due to downstream overflows and plant bypassing.[66]

A project with New York City demonstrated the "duck bill" tide gate.[66] It is a totally passive device, requiring no outside energy to operate and was maintenance-free, yet sealed tightly around large solid objects. Because of its successful demonstration, New York City is planning its installation in other locations, and it is being used in many other municipalities today.

Maintenance

The program has fostered concepts for improved sewerage system inspection and maintenance emphasizing that it is absolutely necessary for a total system approach to municipal water pollution control.

Premature overflows and backwater intrusion during dry as well as wet weather caused by malfunctioning regulators and tide gates, improper diversion settings, and partially blocked interceptors can thus be alleviated. The resulting pollution abatement obtained is a dual benefit of required system maintenance. Some cities have adopted this approach and have gained high CSO control cost benefits.

Infiltration/Inflow (I/I) Control

Various methods to reduce or eliminate I/I, and for infrastructure improvement, have been developed and demonstrated by the program, e.g., inspection,

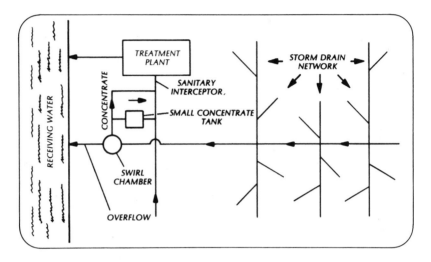

Figure 7.   Urban stormwater runoff pollution control by connecting to existing sanitary sewerage system-schematic.

installation (including trenchless plowing in) and rehabilitation (including liners and Insituform, now used by industry) practices,[67] and new piping (including sulfur impregnation of concrete pipe, which increases pipe strength and corrosion resistance thereby lowering pipe costs and reducing infiltration from deterioration) and jointing materials (including heat shrinkable tubing which expands after installation creating tighter joints).

1.5.3 Storage

Because of the high volume and variability associated with storm flow, storage is considered a necessary control alternative. It is the program's best-documented abatement measure. But it is only the upstream part (process) of the control-treatment system. Project results and theory indicate that storage must be considered at all times in system planning, because it allows for maximum use of existing dry weather plant and downstream drainage facilities, optimum economic sizing of new CSO and stormwater treatment facilities, and results in the lowest cost in terms of pollutant removal.

Storage facilities may be constructed in-line (i.e., flow-through by gravity) or off-line (flow-through by pumping); they may be open or closed; they may be constructed inland and upstream, on the shoreline, or in the receiving water; and they may have auxiliary functions, such as sedimentation treatment, flood protection, flow attenuation to enhance receiving water pollutant assimilation, hazardous materials capture, sewer relief, flow transmission, and dry-weather flow equalization.

A

B

Figure 8.   Prototype rubber "duck bill" tide gate, New York City, NY.

It is important to state that storage facilities can be applied to separate stormwater in the same way they are applied to CSO for bleed or pump back to the sewage treatment plant.

Storage concepts investigated include the conventional concrete holding tanks and earthen basins, and the minimum land requirement concepts of: tunnels, underground and underwater containers, underground "silos", natural and mined under- and aboveground formations, and the use of abandoned facilities and existing sewer lines.[68,69]

The in-receiving water flow balance method (Figure 9) is a recently developed storage alternative.[70] In-receiving water storage facilities contain CSO or stormwater between plastic curtains suspended from floating wooden pontoons. After cessation of the overflow, pumps start and the surrounding water body will enter the compartments and push the storm flow back towards the first compartment where it is pumped to the plant. Thus, the water body is used as a flow balance medium. The pumps will stop based upon receiving sewer and treatment plant handling capacity and an override from a specific ionic sensor (e.g., chlorides) or other parameter sensor that indicates a too-high receiving water dilution.

The storage method is low cost, due to the employment of low cost materials (plastic and wood), the time required to install the unit (several days to months vs. months to a year), and the absence of land requirements. Studies show that costs could be about 5 to 40% of conventional concrete tank costs depending on whether they are installed in inland (relatively quiescent) or marine (relatively rough) water bodies.

The facility that was tested at three locations for stormwater control in Sweden performed very satisfactorily, and was able to take ice and wind loads without adverse impact. It is now being demonstrated in a harsh urban marine site, in Fresh Creek Basin in New York City.

A storage/sedimentation design manual has been completed and we are presently updating it.[71]

1.5.4 Treatment

Due to adverse and intense flow conditions and unpredictable shock-loading effects, it has been difficult to adapt existing treatment methods to storm-generated overflows, especially the microorganism-dependent biological processes. Physical/chemical treatment techniques have shown more promise than biological processes in overcoming storm shock-loading effects. To reduce capital investments, projects have been directed towards high-rate operations approaching maximum loading.

Storm-flow treatment methods demonstrated by the program include physical, physical/chemical, wetlands, biological, and disinfection.[72] These processes, or combinations of these processes, can be adjuncts to the existing sanitary plant or serve as remote satellite facilities at the outfall.

Figure 9.   Isometric view of in-receiving water flow balance method.

Physical/Chemical Treatment

Physical processes with or without chemicals, such as: micro- and fine screens, swirl degritters, high-rate filters (HRF), sedimentation, and dissolved air flotation (DAF), have been successfully demonstrated. Physical processes have shown importance for storm-flow treatment because they are adaptable to automated operation, rapid startup and shutdown, high-rate operation, and resistance to shock loads.

The program thought that the high-rate processes: DAF, micro- and fine-mesh screening, and HRF are ready for municipal installation.

The microstrainer conventionally designed for polishing secondary sewage plant effluent, has successfully been applied to CSOs; and high-rate applications have given SS removals higher than 90%.[73]

Full-scale microscreening units were demonstrated in two locations. In Syracuse, New York, SS removals of about 50% were achieved.[74]

A past Cleveland, Ohio, pilot study using 6-in. columns showed high potential for treating CSOs by a fine screening/HRF system.[75] A large scale (30-in. diameter) fine screening/HRF pilot system was evaluated in New York City for the dual treatment of dry- and wet-weather flows.[76] Removals of SS and BOD were 70% and 40%, respectively. Results from a 5.0 MGD screening and DAF demonstration pilot plant in Milwaukee indicate that greater than 70% removals of BOD and SS are possible.[77,78] By adding chemical coagulants, 85 to 97% phosphate reduction can be achieved as an additional benefit.

Based on program pilot plant studies, two full-scale. screening/DAF proto-type systems (20 and 40 MGD) have been demonstrated.[79] Removals of SS and BOD were 70% and 55%, respectively. Treatment processes, e.g., microscreens and DAF are now being used by municipalities.

The swirl has also been developed for grit removal. The small size, high efficiency and absence of moving parts offer economical and operational advantages over conventional degritting facilities.

A full-scale demonstration of a (16-ft diameter/11 MGD design flow rate) swirl degritter has been completed in Tamwerth, Australia.[80] Removal efficiencies confirmed laboratory results. Compared with a conventional grit chamber, construction costs are halved, and operation and maintenance costs are considerably lower.

Biological Treatment

The biological processes, i.e., trickling filtration, contact stabilization, biodisks, and lagoons, have been demonstrated.[81,82] They have had positive evaluation, but with the exception of long-term storage lagoons, must operate conjunctively with dry-weather flow plants to supply biomass, and require some form of flow equalization.

Disinfection

Because disinfectant and contact demands are great for storm flows,[83] research has centered on high-rate applications by static and mechanical mixing, higher disinfectant concentrations,[84-88] and more rapid oxidants, i.e., chlorine dioxide,[85-88] ozone,[84] and ultraviolet (UV) light; and on-site generation.[84,89,90] Demonstrations in Rochester and Syracuse, New York,[88] East Chicago, Indiana,[91] and Philadelphia, Pennsylvania,[87] indicate that adequate reductions of fecal coliform can be obtained with contact times of two minutes or less by induced mixing and dosing with chlirine and/or chlorine dioxide. A pilot scale UV light demonstration with a contact time of less than ten seconds was conducted at New York City.

The hypochlorite batching facility is still being used in New Orleans, Louisiana, to protect swimming beaches in Lake Poncetrain.[90] The program supported the development of a brine hypochlorite generator now being used in industry.[89]

Treatment/Control Design Guidebook

A compilation of the program's best research efforts in CSO treatment/control over its 18-year duration has been published.[72] Because of flow similarities, this is also an important reference for urban stormwater treatment.

Treatment Process Performance

Treatment process costs and performance in terms of design influx (gpm/ft^2) and BOD and SS removal efficiency is provided in Table 7. The high-rate performance of the swirl, microstrainer, screening/HRF, and screening/DAF systems is apparent when compared to sedimentation.

Maximizing Treatment

The operator may maximize wastewater treatment at the sanitary plant during wet weather, and should try to contain as much flow or treat as much wastewater as possible during a storm-flow occurrence. Treatment maximization can be enhanced by advanced signals of relatively high storm flows from remotely stationed rain gauges and/or radar. This would serve to reduce wet-weather bypassing, which at the beginning of storm flow can have a high pollutant concentration (including floatables), as previously described. Although this extra plant burden may decrease treatment efficiencies somewhat and create added sludge or solids handling problems, these practices for only short periods during storm flows are well worth the effort. If the operator determines that the hydraulic loading will cause a serious upset of a unit process, then primary treatment plus disinfection should be considered as a minimum measure. Effectiveness evaluation should be made for the entire treatment plant drainage area and be based on the total mass of pollutants captured or prevented from overflowing the combined sewers while also taking into account settling tank efficiency decrease as a function of higher influent hydraulic loading (or overflow rates).

1.5.5 Sludge/Solids

Another program area is the sludge and solids associated with storm-flow treatment. Sludge handling and disposal must be considered an integral part of CSO treatment because it significantly affects the efficiency and cost of the total waste treatment system. Studies have shown that the annual quantity of CSO solids is at least equal to solids from dry-weather flow.[92] This is a significant finding for municipal pollution control programs. The results of a project on CSO sludges are covered in three published reports covering characterization impact,[92] assessment,[93] and treatability.[94] A similar study was conducted for separate stormwater sludge and residuals.[95]

1.5.6 Integrated Systems

The most promising and common approach to urban storm flow management involves the integration of control and treatment. Integrated systems is divided

Table 7. Wet-Weather Treatment Plant Performance Data

Device	Control Alternatives	Design Loading Rate (gpm/ft^2)	Removal Efficiency (%)	
			BOD$_5$	SS
Primary	Swirl concentrator	60	25–60	50
	Microstrainer	20	40–60	70
	High-rate filtration	24	60–80	90
	Dissolved air flotation	2.5	50–60	80
	Sedimentation	0.5	25–40	55
	Representative performance		40	60
Secondary	Contact stabilization		75–88	90
	Physical-chemical		85–95	95
	Representative performance		85	95

into storage/treatment, dual-use wet-weather-flow/dry-weather-flow facilities, and control/treatment/reuse.

Storage/Treatment

When there is storage, there is treatment by settling, pump-back/ bleed-back to the municipal works, and sometimes disinfection. Treatment which receives detention also provides storage. The break-even economics of supplying storage must be evaluated when treatment is considered. The program has demonstrated all of these storage/treatment concepts full scale.

Dual Use Wet-Weather-Flow/Dry-Weather-Flow Facilities

The concept of dual use is maximum utilization of wet-weather facilities during nonstorm periods and maximum utilization of dry-weather facilities during storm flows. The program has demonstrated the full scale dual-use of high-rate trickling filters,[96] contact stabilization,[81,82] HRF,[76] and equalization basins.[97] Various municipalities are employing dual-use microscreening.

In Clatskanie, Oregon, a full-scale dual facility constructed to alleviate flow bypassing caused by excessive infiltration was evaluated. The plant is in permanent use. Both wet- and dry-weather flow treatment is provided for in the same units and consists of primary sedimentation and conventional activated sludge for dry-weather periods converting to higher rate DAF and contact stabilization for wet-weather periods.[81]

Control/Treatment/Reuse

"Control/Treatment/Reuse" is a "catch-all" for all integrated systems. A prime consideration should be the various nonstructural and land management techniques. In Mt. Clemens, Michigan, a series of three "lakelets" have been incorporated into a CSO treatment/park development.[98] Treatment is being provided so that these lakes are aesthetically pleasing and allow for recreation and reuse for irrigation.

An inhouse paper covering subpotable reuse was published by the ASCE.[99]

Wetlands

The use of wetlands for urban runoff pollution control has been investigated.[100] It has been found, that with controlled runoff entry and wetlands management and maintenance, significant receiving-lake water benefits are obtained without degrading the wetlands.[101]

1.6 CONCLUSIONS

In general, on a mass basis, toxics, bacteria, oxygen-demanding, suspended, and visual matter in CSO and urban stormwater are significant. Ignoring the problem because it seems to be too costly to solve, will not make the problem go away. The integrated approach to wet-weather pollution control is the only way that is going to be feasible, economical and, therefore, acceptable. Potentially tremendous "bangs for the bucks" can be derived from wet-weather pollution control research fostering integrated solutions. As you can see, the SCSP has investigated a problem, proven its significance, and developed a gamut of design and control techniques that has led our nation and been accepted internationally. Better advantage needs to be taken of proven technology.

And as was discussed, and because of the hundreds of millions of dollars being spent annually, much more research still needs to be done.

1.7 RECOMMENDATIONS FOR THE FUTURE

1.7.1 Receiving-Water Impacts

Ties between receiving-water quality and storm-flow discharges must be clearly established and delineated. Quantification of the impairment of beneficial uses and water quality by such discharges is a major goal. Project results indicate the potential for significant impact to receiving waters of wet-weather flows. Control of runoff pollution can be a viable alternative for maintaining receiving water quality standards. However, the problems found seem to be site-specific in nature. Therefore, site-specific surveys are required that must consider the effects of larger materials and floatables near the outfalls (the nearfield). Based on results from these surveys, control may be warranted.

1.7.2 Indicator Microorganisms/Disinfection Requirements and Technology

As discussed earlier, research is warranted for finding better indicator microorganisms for the disease-causing potential of CSO and stormwater, associated disinfection requirements, and disinfection technology since disinfection costs will be great for the high storm flows encountered.

1.7.3 Toxics Characterization/Problem Assessment/Control-Treatment

Results from a limited in-house effort, and EPA OWRS studies (including the Nationwide Urban Runoff Program [NURP] study) indicate that urban stormwater runoff and CSO contain significant quantities of toxic substances (priority pollutants). Without toxic and industrial runoff problem assessment and control, our various hazardous substances cleanup and control programs (under CERCLA/SARA, RECRA, TSCA, etc.) may be done in vain. Additional investigation of the significance of concentrations and quantities of toxic pollutants with regard to their health effects or potential health effects and ecosystem effects is required. A need exists to evaluate the removal capacity of conventional and alternative treatment technologies and BMP's for these toxics and to compare their effectiveness with estimated removal needs to meet water quality goals. From this comparison further advanced treatment and control for toxic substances will need to be developed.

1.7.4 Industrial Stormwater Runoff Problem Assessment/Control

Permits for industrial stormwater runoff along with follow-up compliance and control are now mandated requirements (WQA Section 405 and CWA Section 402 (p)). There are thousands of industrial sites in the country with pollutants and toxicants in their runoff. Research and development for problem assessment and control of industrial stormwater runoff are needed to support these mandates; especially because research has never been done in this area.

1.7.5 Moratorium Sources Runoff Problem Assessment/Control

Research support is required for the assessment and control of storm-water runoff from all moratorium sources (i.e., municipalities with populations less than 100,000 and commercial/institutional areas) as mandated by the WQA Section 405.

1.7.6 Sewer System Cross-Connections

Investigations have shown that sanitary and industrial contamination of separate-storm sewers (by cross-connections) is a nationwide problem. In other words, a significant number of separate stormwater drainage systems function as combined sewer systems. Therefore, a nationwide effort on both federal and local levels to alleviate the pollution impacts from discharges of these systems is required. It may be better to classify such deteriorated drainage systems as combined systems for pollution control purposes and priorities. More research on detection and control is needed because of the large sums of money that will be spent on corrective action.

1.7.8 Leaking Underground Storage Tanks (UST)

Many leaks from UST enter utility trenches and lines, e.g., sewer networks. Pollution abatement costs would be significantly lower if methodologies are developed to enable municipalities to detect and control UST leaks via these utility systems.

1.7.9 Integrated Stormwater Management

The most effective solution methodology for wet-weather pollution problems must consider (1) wet-weather pollution impacts in lieu of blindly upgrading existing municipal plants, (2) structural vs. nonstructural techniques, (3) integrating dry- and wet-weather flow systems/control to make maximum use of the previously existing sewerage/drainage systems during wet conditions and maximum use of wet-weather control/treatment facilities during dry weather, and (4) the segment or bend on the percent pollutant control vs. cost curve in which cost differences accelerate at much higher rates than pollutant control increases, although load discharge or receiving-water requirements will dictate, ultimately, the degree of control/treatment required.

Flood and erosion control technology must be integrated with pollution control, so that the retention and drainage facilities required for flood and erosion control can be simultaneously designed or retrofitted for pollution control. Upstream storage should also be designed to lessen size and cost requirements for downstream drainage. If land management and nonstructural/low-structurally intensive techniques are maximized and integrated, there will be less to pay for the extraction of pollutants from storm flows in the potentially more costly downstream plants. There is a significant need to further develop and demonstrate various forms of integrated stormwater management.

1.7.10 New and Innovative Stormwater Control

New research and development must be devoted to the low-cost separate

stormwater pollution control concepts, e.g., swirls and smaller storage units for
bleed-back to the existing dry-weather plant.

1.7.11 Surface and Groundwater Interfacing

Surface and groundwater have never been interfaced in the area of pollutant
routing. Runoff problems cannot be adequately assessed without this interface. For example, enhancing surface runoff infiltration to groundwater by
applying certain BMPS, i.e., porous pavement may cause a groundwater
pollution problem that in turn may create a surface water pollution problem
later.

1.7.12 Landfill and Waste Site Runoff Control

Landfill and waste site runoff/leachate convey vast quantities of toxic and
other pollution substances to surface and groundwater. Pollutant routing and
control technologies should be developed.

1.7.13 Institutional/Socio/Economic Conflicts

Some of the most promising opportunities for cost-effective environmental
control are multipurpose in nature. However, there are institutional problems
that hinder their implementation. First, the autonomous federal and local
agencies and professions involved in flood and erosion control, pollution
control, and land management and environmental planning must be integrated
at both the planning and operation levels. Multiagency incentives (e.g., grant
coverage) and rules must be adequate to stimulate such an approach. For
example, the EPA would have to join with the Corps of Engineers, Soil
Conservation Service, Department of Transportation, and perhaps other federal
agencies as well as departments of pollution control, sanitation, planning, and
flood control at the local level.

Another problem is that construction grant (and other) incentives are geared
towards structurally intensive projects which may counter research findings in
the area of optimal solutions. Optimized wet-weather pollution involves a city-
wide approach including the integration of structural as well as low-structural
controls. The low-structural measures are more labor intensive. Construction
grant funding does not presently address this expense and accordingly, municipalities are discouraged from using them.

REFERENCES

1. Lager, J.A., and W.G. Smith. "Urban stormwater management and technology: an assessment," U.S. EPA Report-670/2-74-040, NTIS No. PB
240 687 (1974).

2. Lager, J.A., W.G. Smith, W.G. Lynard, R.M. Finn, and E.J. Finnemore. "Urban stormwater management and technology: update and user's guide," "U.S. EPA Report-600/8-77-014, NTIS No. PB 275 654 (1977).
3. Sullivan, R.H. "Nationwide evaluation of combined sewer overflows and urban stormwater discharges, Vol. III — characterization of discharges," U.S. EPA Report-600/2-77-064c, NTIS No. PB 272 107 (1977).
4. Huber, W.C., et. al. "Urban rainfall-runoff-quality data base," U.S. EPA Report-600/2-81-238, NTIS No. PB 82-221 094 (1981).
5. Spiegel, S.J., et al. "Evaluation of urban runoff and combined sewer overflow mutagenicity," U.S. EPA Report-600/2-84-116, NTIS No. PB 84-211 168 (1984).
6. Olivieri, V.P., C.W, Kruse, K. Kawata and J.E. Smith. "Microorganisms in urban stormwater," U.S. EPA Report-600/2-77-087, NTIS No. PB 272 245 (1977).
7. "Urban stormwater and combined sewer overflow impact on receiving water bodies," in Proceedings of National Conference Orlando, Florida, U.S. EPA Report-600/9-80-056, NTIS No. PB 81-155 6 (November 26–28, 1979).
8. Field, R., and R. Turkeltaub. "Urban runoff receiving water impacts: program overview,"J. Environ. Eng. Div. ASCE 107: 83 (1981).
9. Sullivan, R.H., M.J. Manning, J.P. Heaney, M.A. Medina, Jr., M.P. Murphy, S.J. Nix and S.M. Hasen. "Nationwide evaluation of combined sewer overflows and urban stormwater discharges, Vol. I — executive summary," U.S. EPA Report-600/2-77-064a, NTIS No. PB 273 133 (1977).
10. Heaney, J.F., W.C. Huber, M.A. Medina, Jr., M.P. Murphy, S.J. Nix and S.M. Hasen. "Nationwide evaluation of combined sewer overflows and urban stormwater discharges, Vol. II — cost assessment and impacts," U.S. EPA Report-600/2-77-064b, NTIS No. PB 266 005 (1977).
11. Pitt, R., and M. Bozemen. "Sources of urban runoff pollution and its effects on an urban creek," U.S. EPA Report-600/S2-82-090, NTIS No. PB 830111 021 (1982).
12. Tomlinson, R.D., et al. "Fate and effects of particulates discharge by combined sewers and storm drains," U.S. EPA Report-600/2-80-111, NTIS No. PB 81-118 390 (1980).
13. Keefer, N., et al. "Dissolved oxygen impact from urban storm runoff," U.S. EPA Report-600/2-79-156 (1979).
14. Meinholz, T.L., et al. "Verification of the water quality impacts of combined sewer overflows," U.S. EPA Report-600/2-79-155 (1979).
15. Medina, M. "Level III: Receiving water quality modeling for urban stormwater management," U.S. EPA Report-600/2-79-100, NTIS No. PB 80-134 406 (1979).
16. Medina, M.A., et al. "River quality model for urban stormwater impacts," U.S. EPA Report-600/J-81-234, NTIS No. PB 81-221 087 (1981).

17. Moffa, P.E., et al. "Methodology for evaluating the impact and abatement of combined sewer overflows: a case study of Onondaga Lake, New York," U.S. EPA Report-600/8-80-048, NTIS No. PB 81-141 913 (1980).
18. Mancini, J.L. "Development of methods to define water quality effects of urban runoff," Project Summary U.S. EPA Report-600/S2-83-125, NTIS No. PB 84-122 928 (1983).
19. Mancini, J.L. "A method for calculating effects on aquatic organisms, of time varying concentrations," *Water Res.* 17: 1355 (1983).
20. Foreman, K.M. "Field testing of prototype acoustic flowmeter," U.S. EPA Report-600/2-79-084, NTIS No. PB 80-121 544 (1979).
21. Anderson, R.J., and S.S. Bell. "Wastewater flow measurement in sewers using ultrasound," U.S. EPA Report- 600/2-76-243, NTIS No. PB 262 902 (1976).
22. Shelley, P.E. "Design and testing of prototype automatic sewer sampling system," U.S. EPA Report-600/2-76-006, NTIS No. PB 252 613 (1976).
23. Shelley, P.E., and G.A. Kirkpatrick. "Sewer flow measurement - a state-of-the-art assessment." U.S. EPA Report-600/2-75-027, NTIS No. PB 250 371 (1975).
24. Shelley, P.E., and G.A. Kirkpatrick. "An assessment of automatic sewer flow samplers-1975," U.S. EPA Report-600/2-75-065, NTIS No. PB 250 987 (1975).
25. Huber, W.C., et al. "Storm water management model user's manual, version III," U.S. EPA Report pending, (interim availability: University of Florida).
26. Roesner, L.A., et al. "Storm water management model user's manual, version III, addendum I — extran," U.S. EPA Report pending, (interim availability: University of Florida).
27. Geiger, W.F., and H.R. Dorsch. "Quantity-quality simulation (QQS): a detailed continuous planning model for urban runoff control, Vol. I — model description, testing, and applications," U.S. EPA Report-600/2-80-011, NTIS No. PB 80-190 507 (1980).
28. Geiger, W.F., and H.R. Dorsch. "Quantity-quality simulation (QQS): a detailed continuous planning model for urban runoff control, Vol. II —user's manual," U.S. EPA Report-600/2-80-116, NTIS No. PB 80-221 872 (1980).
29. Litwin, Y.J., et al. "Areawide stormwater pollution analysis with the macroscopic planning (ABMAC) model," Project Summary U.S. EPA Report-600/S2-81-223, NTIS No. PB 82-107 947 (1981).
30. Smith, W.G., and M.E. Strickfaden. "EPA macroscopic planning model (EPAMAC) for stormwater and combined sewer overflow control: application guide and user's manual," U.S. EPA Report-600/2-83-086, NTIS No. PB 83-259 689 (1983).
31. Watt, T.R., et al. "Sewerage system monitoring and remote control," U.S. EPA Report-670/2-75-020, NTIS No. PB 2 107 (1975).

32. St. Paul Minnesota Metropolitan Sewer Board. "Dispatching systems for control of combined sewer losses," U.S. EPA Report-11020FAQ03/71, NTIS No. PB 203 678 (1971).

33. Leiser, C.P. "Computer management of a combined sewer system," U.S. EPA Report-670/2-74-022, NTIS No. PB 235 717 (1977).

34. Lager, J.A., W.G. Smith, W.G. Lynard, R.M. Finn and E.J. Finnemore. "Urban stormwater management and technology: update and user's guide," U.S. EPA Report-600/8-77-014, NTIS No. PB 275 654 (1977).

35. Lager, J.A., and W.G. Smith. "Urban stormwater management and technology: an assessment," U.S. EPA Report-670/2-74-040, NTIS No. PB 240 687 (1977).

36. Brater, E.F., and J.D. Sherrill. "Rainfall-runoff relations on urban and rural areas," U.S. EPA Report-670/2-75-046, NTIS No. PB 2 830 (1975).

37. Yen, B.C., and V.T. Chow. "Urban stormwater and runoff determination of volumes and flowrates," U.S. EPA Report-600/2-76-116, NTIS No. PB 253 410 (1976).

38. Field, R., et al. "Porous pavement: research, development, and demonstration," *Transp. Eng. J. ASCE* 108: 244 (1982).

39. Diniz, E. "Porous pavement: phase I — design and operational criteria," U.S. EPA Report-600/2-80-135, NTIS No. PB 81-104 796 (1980).

40. Goforth, G.F., E.V. Diniz and J.B. Rauhut. "Stormwater hydrological characteristics of porous and conventional paving systems," U.S. EPA Report-600/2-83-106, NTIS No. PB 84-123 728 (1983).

41. Benjes, H.H., Jr. "Cost estimating manual — combined sewer overflow storage treatment," U.S. EPA Report-600/2-76-286, NTIS No. PB 266 359 (1976).

42. Benjes, H.H., Jr., and R. Field. "Estimate sewer overflow facility costs," *Water Wastes Eng.* 56 (1978).

43. Sonnen, M. "Abatement of deposition and scour in sewers," U.S. EPA Report-600/2-77-212, NTIS No. PB 276 585 (1977).

44. Yen, B.C., et al. "Stormwater runoff on urban areas of steep slope," U.S. EPA Report-600/2/77-168, NTIS No. PB 272 755 (1977).

45. Kaufman, H.L., and F.H. Lai. "Conventional and advanced sewer design concepts for dual purpose flood and pollution control — a preliminary case study, Elizabeth, NJ," U.S. EPA Report-600/2-78-090, NTIS No. PB 285 663 (1978).

46. Field, R., and D.A. Weisman. "A planning and design guidebook for combined sewer overflow control," U.S. EPA Report-600/2-82-084, NTIS No. PB 82-259 235 (1982).

47. Wuelschleger, R.E., et al., "Methodology for the study of urban storm generated pollution and control," U.S. EPA Report-600/2-76-145, NTIS No. PB 258 243 (1976).

48. Dayrymple, R.J., et al. "Physical and settling characteristics of particulates in storm and sanitary wastewater," U.S. EPA Report-670/2-75-011, NTIS No. PB 2 001 (1975).

49. Field, R. "Stormwater pollution control: a new technology," General Services Administration (GSA), National Audio Visual Center, Washington, DC.

50. Lynard, W., et al. "Urban stormwater management and technology: case histories," U.S. EPA Report-600/8-80-035, NTIS No. PB 81-107 153 (1980).

51. Murphy, C. B., T.J. Quinn and J.E. Stewart. "Best management practices: implementation," U.S. EPA Report-905/9-81-002, NTIS No. PB 82-169 210 (1981).

52. Pitt, R.E. "Demonstration of nonpoint pollution abatement through improved street cleaning practices," U.S. EPA Report-600/2-79-161, NTIS No. PB 80-108 988 (1979).

53. Pitt, R. E. "Characterization, sources, and control of urban runoff by street and sewerage cleaning," Project Summary pending.

54. Aronson, G.L., et al. "Evaluation of catchbasin performance for urban stormwater pollution control," U.S. EPA Report-600/2-83-043, NTIS No. PB 83-217 745 (1983).

55. Pisano, W. C., and C.S. Queriroz. "Procedures for estimating dry weather pollutant deposition in sewerage systems," U.S. EPA Report-600/2-77-120, NTIS No. PB 270 695 (1977).

56. Pisano, W.C., et al. "Dry-weather deposition and flushing for combined sewer overflow pollution control," U.S. EPA Report-600/2-79-133, NTIS No. PB 80-118 524 (1979).

57. Kaufman, H.L., and F.H. Lai. "Review of alternatives for evaluation of sewer flushing Dorchester area — Boston," U.S. EPA Report-600/2-80-118, NTIS No. PB 81-1 648 (1980).

58. Chandler, R.W., and W.R. Lewis. "Control of sewer overflows by polymer injection," U.S. EPA Report-600/2-77-189, NTIS No. PB 272 654 (1977).

59. American Public Works Association. "Combined sewer regulator overflow facilities," U.S. EPA Report-11022DMU07/70 (1970).

60. American Public Works Association. "Combined sewer regulation and management: a manual of practice," U.S. EPA Report-11022DMU08/70, NTIS No. PB 195 676 (1970).

61. Freeman, P.A. "Evaluation of fluidic combined sewer regulators under municipal service conditions," U.S. EPA Report-600/2-77-071, NTIS PB 272 834 (1977).

62. Field, R. "The dual-functioning swirl combined sewer overflow regulator/concentrator," U.S. EPA Report-670/2-73-059, NTIS No. PB 227 182 (1973).

63. Field, R., and H.E. Masters. "Swirl device for regulating and treating combined sewer overflow," U.S. EPA Report-625/2-77-012 (1977).

64. Sullivan, R.H., et al. "Swirl and helical bend pollution control devices: design manual," U.S. EPA Report-600/8-82-013, NTIS No. PB 82-266 172 (1982).

65. Matthews, T.M., et al. "Hydrobrakes regulated storage system for stormwater management," U.S. EPA Project Summary Report-600-/S2-83-097, NTIS No. PB 84-110 378 (1983).

66. Field, R. "An overview of the U.S. Environmental Agency's storm and combined sewer program collection system research," *Water Res.* 16: 859 (1982).

67. Driver, T., and M. Olsen. "Demonstration of sewer relining by the in situ form process, Northbrook, IL," U.S. EPA Report-600/2-83-064, NTIS No. PB 83-245 878 (1983).

68. Field, R., and E.J. Struzeski. "Management and control of combined sewer overflows," *J. Water Pollut. Control Fed.* 44: 1393 (1972).

69. Field, R., and J.A. Lager. "Urban runoff pollution control state-of-the-art," *J. Environ. Eng. Div. ASCE* 101: 107 (1975).

70. Soderland, H., and Kjessler & Mannerstrale AB. "Flow balancing method for stormwater and combined sewer overflow," Swedish Council for Building Research, Stockholm, Sweden, 1982, p. 1.

71. Smith, William G., et al. "Storage/sedimentation facilities for control of storm and combined sewer overflows design manual," U.S. EPA Report pending.

72. Field, R., and D.A. Weisman. "A planning and design guidebook for combined sewer overflow control and treatment," U.S. EPA Report-600/2-82-084, NTIS No. PB 82-259 235 (1982).

73. Maher, M. B. "Microstraining and disinfection of combined sewer overflows — phase III," U.S. EPA Report-670/2-74-049, NTIS No. PB 235 771 (1974).

74. Drehwing, F., A.J. Oliver, D.A. MacArthur and P.E. Moffa. "Disinfection treatment of combined sewer overflows, Syracuse, New York," U.S. EPA Report-600/2-79-134, NTIS No. PB 80-113 459 (1979).

75. Nebolsine, R., et al. "High-rate filtration of combined sewer overflows," U.S. EPA Report-11023EY104/72, NTIS No. 211 114 (1972).

76. Innerfield, H., and A. Forndran. "Dual process high-rate filtration of raw sanitary sewage and combined sewer overflows," U.S. EPA Report-600/2-79-015, NTIS No. PB 296 626/AS (1979).

77. Gupta, M.K., et al. "Screening/flotation treatment of combined sewer overflows, Vol. I Bench scale and pilot plant investigations," U.S. EPA Report-600/2-77-069a, NTIS No. PB 272 834 (1977).

78. Rex Chainbelt, Inc. Ecology Division. "Screening/floatation treatment of combined sewer overflows," U.S. EPA Report-11020FDC-01/72 (1972)

79. Meinholz, T.L. "Screening/floatation treatment of combined sewer overflows, Vol. II — Full scale operation, Racine, WI," U.S. EPA Report-600/2-79-106a, NTIS No. PB 80-130 693 (1979).

80. Shelley, G. J., et al. "Field evaluation of a swirl degritter at Tamworth N.S.W., Australia," U.S. EPA Report-600/2 81-063, NTIS No. PB 81-187 247 (1981).

81. Benedict, A.H., and V.L. Roelfs. Joint dry-wet weather treatment of municipal wastewater at Clatskanie, Oregon, U.S. EPA Report-600/2-81-061, NTIS No. PB 81-187 262 (1981).

82. Agnew, R.W., et al. "Biological treatment of combined sewer overflow at Kenosha, WI," U.S. EPA Report-670/2-75-019, NTIS No. PB 2 107 9 (1975).

83. Field, R., V.P. Olivieri, E.M. Davis, J.E. Smith and E.C. Tifft, Jr. "Proceedings of workshop on microorganisms in urban stormwater," U.S. EPA Report-600/2-76-244, NTIS No. PB 263 030 (1973).

84. Glover, G.E., and G.R. Herbert. "Micro-straining and disinfection of combined sewer overflows — phase II," U.S. EPA Report-R2-73-124, NTIS No. PB 219 879 (1973).

85. Drehwing, F.J., et al. "Combined sewer overflow abatement program, Rochester, NY, Vol. II. Pilot plant evaluations," U.S. EPA Report-600/2-79-031b, NTIS No. PB 80-159 262 (1979).

86. Moffa, P.E., E.C. Tifft, Jr., S.L. Richardson and J.E. Smith. "Bench-scale high-rate disinfection of combined sewer overflows with chlorine and chlorine dioxide," U.S. EPA Report-670/2-75-021. NTIS No. PB 2 296 (1975).

87. Mahre, M.B. "Microstraining and disinfection of combined sewer over-flows — phase III," U.S. EPA Report-670/2-74-049, NTIS No. PB 235 771 (1974).

88. Drehwing, F., et al. "Disinfection/treatment of combined sewer overflows, Syracuse, New York," U.S. EPA Report-600/2-79-134, NTIS No. PB 80-113 459 (1972).

89. Leitz, F.B., et al. "Hypochlorite generator for treatment of combined sewer overflows," U.S. EPA Report-11023DAA03/72, NTIS No. PB 211 243 (1972).

90. Pontius, U.R., et al. "Hypochlorination of polluted stormwater pumpage at New Orleans," U.S. EPA Report-670/2-73-067, NTIS No. PB 228 581 (1973).

91. Connick, D.J., et al. "Evaluation of a treatment lagoon for combined sewer overflow," U.S. EPA Report-600/S2-81-196, NTIS No. PB 82-105 214 (1981).

92. Gupta, M.K., E. Bollinger, S. Vanderah, C. Hansen and M. Clark. "Handling and disposal of sludges from combined sewer overflow treat-ment — phase I (characterization)," U.S. EPA Report-600/2-77-053a, NTIS No. PB 270 212 (1977).

93. Huibregtse, K.R., G.R. Morris, A. Geinopolos and M.J. Clark. "Handling and disposal of sludges from combined sewer overflow treatment — phase II (impact assessment)," U.S. EPA Report-600/2-77-053b, NTIS No. PB 280 309 (1977).

94. Osantowski, R., A. Geinopolos, R.E. Wullschleger and M.J. Clark. "Handling and disposal of sludges from combined sewer overflow treatment — phase III (treatability studies)," U.S. EPA Report-600/2-77-053c, NTIS No. PB 281 006 (1977).

95. Huibregtse, K.R., and A. Geinopolos. "Evaluation of secondary impacts of urban runoff pollution control," U.S. EPA Report-600/2-82-045, NTIS No. PB 82-230 319 (1982).

96. Homack, P., et al. "Utilization of trickling filters for dual treatment of dry and wet weather flows," U.S. EPA Report-670/2-73-071, NTIS No. PB 231 251 (1973).

97. Welborn, H.L. "Surge facility for wet- and dry-weather flow control," U.S. EPA Report-670/2-74-075, NTIS No. PB 283 905 (1974).

98. Mahida, V.U., and F.J. DeDecker. "Multi-purpose combined sewer overflow treatment facility, Mt. Clemens, MI," U.S. EPA Report-670/2-75-010, NTIS No. PB 2 914 (1975).

99. Field, R., and C-Y. Fan. "Industrial reuse of urban stormwater," *J. Environ. Eng. Div. ASCE* 107: 171 (1981).

100. Litwin, Y.J., et al. "The use of wetlands for water pollution control," U.S. EPA Report-600/2-82-086, NTIS No. PB 83-107 466 (1982).

101. Hickock, et al. "Urban runoff treatment methods, Vol. I — Non-structural wetland treatment," U.S. EPA Report-600/2-77-217, NTIS No. PB 278 172 (1977).

2 INTEGRATED WATER MANAGEMENT BACKGROUND TO MODERN APPROACH, WITH TWO CASE EXAMPLES

2.1 INTRODUCTION

Most of the present scientific efforts and technical developments within water supply, water treatment technology, environmental protection and restoration methods are still devoted to technology designed by, and applicable in rich, developed countries. These technologies emerged in a long process of development during the last century, mainly in European countries and in North America. During this period, the level of science and technology, as well as our understanding of interconnections between human actions and their environmental consequences, have changed drastically. Thus, the existing infrastructure in developed countries is an effect of an evolutionary process of changing approaches, which is a "trial and error process". Departing from our present level of understanding of cause-effect relationships, we realize that our rural and urban infrastructures are not functioning satisfactorily in a global environmental sense.

Changes in the approach to water management and environment protection has recently occurred in developed countries. During the recent decade at least three major changes in our perception of the world have occurred:

1. We have realized that the environmental problems which are the result of human activities, are global.
2. We have realized that environmental degradation, especially in developing countries, is continuing in spite of great efforts.
3. Eastern Europe, which joined the world community of free countries, brought frightening news about environmental negligence and degradation.

0-87371-805-4/93/$0.00+$.50
© 1993 by Lewis Publishers, Inc.

Moreover, we have realized that water management can only be efficient, in terms of satisfying human needs and environmental constraints, if it is integrated with the management of other resources and human activities. This newly-gained perception of the world has brought desolation and mistrust, because it showed that our previous optimistic, monodisciplinary and solely-technical approach was perhaps not so wise. Even highly developed countries with expensive infrastructure and sophisticated water works and treatment plants still experience various problems in satisfying basic human needs with respect to the supply of clean drinking water, sound wastewater and waste disposal, maintenance and reconstruction of existing facilities, etc.

Cities of industrialized countries still contribute to the continuing degradation of the global environment, despite all efforts and the use of advanced sewerage systems and sophisticated treatment facilities. Water treatment plants do not eliminate the most hazardous substances. Stormwater delivers heavy metals and other nondegradable toxic elements to the environment. Discharges from industrial and construction activities cause chemical contamination of surface waters and aquifers. Nature's revenge takes the form of forest death, extinction of species, decertification, pollution of previously untouched areas, climate change, etc. Changes in the approach to water management have not yet occurred in developing countries and Eastern Europe. Centralized plans of water resources management and environmental protection actions still proceed (with few exceptions) along the traditional lines of monodisciplinary, large-scale-thinking. "End of pipe" approaches are used instead of source control and local disposal, i.e., reaction instead of prevention, effluent dilution instead of selective concentration and reuse. The present economic situation in many countries stops or delays even these plans.[1,2] We should act in order to accelerate a necessary change of paradigm in these countries.

Problems with management of water resources and environmental protection are different in various parts of the world. In developing countries infrastructure is usually inadequately developed or lacking; environmental problems are rapidly increasing, especially in large cities; basic needs are seldom satisfied; and economies do not allow fast improvements. In this situation, an ecological and global approach is ranked low in priority. For many developing countries, new solutions must be found in order to stop environmental degradation. These new solutions should be based on an ecological approach and environmental concern, and they should also be economically effective, because only such solutions are affordable for these countries. Due to climatic, geological, socio-economic and other differences, certainly different solutions will be needed for various parts of the world.

The environmental pollution generated by stormwater releases may seem small in comparison with pollution due to the discharge of untreated wastewater and industrial effluents. However, the more wastewater is treated, the higher the relative importance of stormwater releases. While focusing on environmental problems, we should not forget other disturbances in the social functions of a city. Several vital functions of a city and the surrounding environment are disturbed by stormwater. For example, the flooding of streets causes traffic problems, the flooding of houses and basements brings economical losses, the performance of treatment plants is disturbed during the rain, and wash-off from urban surfaces increases the risk of receiving-water contamination. Thus, stormwater, as well as other types of water must be considered in the process of water management. On the other hand, integrated water management in cities should be based on our present level of problem comprehension, i.e., the present technical paradigm.

2.2 DEVELOPMENT OF A NEW APPROACH

The process of departing from the traditional, monodisciplinary approach is currently going on in the developed world. Goals of water management have already shifted from the traditional storm drainage and "end of pipe" approaches towards sustainable ecological solutions more in peace with nature. The new approach is based on a deeper understanding of the cyclicity of material and energy flows in nature. In searching for new technologies, we should leave the traditional ways of thinking within our professional areas and create links to other related disciplines. Ecology and ecological engineering are examples of sciences that can contribute much to the development of new thinking and new technical solutions. There are certain key expressions which represent this modern approach. These expressions also represent our present stage of knowledge and they are valid across all national, climatic, and regional borders. They should be applied independently at the present stage of development in any country. The technologies that are appropriate for different regions may emerge from the application of the following expressions:

- *Integrated system approach* with both structural and nonstructural elements, in contrast to a narrow-minded, solely-technical approach.
- *Small scale*, in contrast to technological large-scale thinking; it is less expensive to construct small-scale treatment units than large ones. The sources of wastewater are closer, plants are less vulnerable and it is possible to design plants adjusted to the local needs and conditions.
- *Source control* instead of the "end of pipe" approach;[3] it is less expensive to reduce stormwater volumes at the source than construct huge conduits.
- *Local disposal and reuse* instead of exploitation and wastefulness. Volumes of stormwater, wastewater, and wastes may be reduced at the source by changing the routines in the production stage or/and by local disposal. Water may be reused for industry after local treatment; wastes may be reused after separation at the source; ashes may be used for road construction; heavy metals can be extracted from effluents using bacterial uptake, etc. Traditional solutions, both for treating wastewater and solids, mix together several pollution components which makes reuse difficult. Sepa-

ration on the production stage is one possibility, another is to use biological systems which act selectively.

• *Pollution prevention* instead of reacting to damages.

• *The ecological approach and the use of biological systems* in wastewater, stormwater, and solid waste management. Biological systems may complement or substitute traditional treatment facilities. For example, biological activity in the upper part of soil may create similar physico-chemical conditions as in a best treatment plant. Bacteria, algae, plants, and animals may together constitute a system which selectively removes and concentrates all pollution components in any type of water. Polluted water enters the system, clean water leaves the system *plus* we have re-usable resources such as wood (energy), paper, livestock food, chemicals, etc. Existing or artificially created ecological systems may be used in the treatment of water. Wetlands, ecotones close to the rivers, and riparian zones with root uptake may take part in the purification of water. The use of wetlands and plant filters for the treatment of polluted waters may became a major method for use by developing countries where construction of traditional treatment plants is impossible due to economical reasons. The use of plant filters in conjunction with the production of renewable biomass for use as fuel is equally attractive for developed countries where alternative energy sources must be developed in order to reduce CO_2 emission. This method is also an alternative to the development of new steps in water treatment plants for the reduction of nitrogen emissions, and it may also be an attractive method for the reduction of pollution from stormwater. These ecologically sound and economically attractive methods may soon replace traditional treatment methods. Thus, wastewater may become a valuable resource which is used in the biological process of biomass production, a process close to the natural biological cycles of matter and energy.[4]

The development and improvement of biological systems which may be used in alternative treatment of polluted water is at present a field of intensive work for many scientists in various disciplines.[5-7] Such systems must be further developed before they can really replace traditional treatment methods on a larger scale; however, parts of them are in use today and the number of applications increases rapidly.

2.3 APPLICATION EXAMPLE: SWEDEN

In practice, 100% of all communal waters are treated in three-stage plants in Sweden. In this situation, the relative importance of pollution generated by surface runoff, combined sewer overflows, and the agriculture is increasing. Thus, stormwater enhancement with respect to quantity and quality has became a desirable goal for city planners and practitioners dealing with source control methods in Sweden.[8] The main goal is to relieve overloaded sewer systems and to reduce combined sewer overflows. The separation of waste- and stormwater systems was abandoned in the early 1980s because it was found economically unacceptable and technically insufficient to prevent a further degradation of receiving water quality.[9] The total pollution loads from a city would be hardly reduced and the risk of toxic effects on ecologic systems in rivers during heavy rainfalls could be increased.[10] The inlet controls used in Sweden mainly consist of various kinds of infiltration and percolation facilities. Several hundreds of such facilities have been installed.

Between 1981 and 1983, more than 14,000 wastewater infiltration facilities have been constructed for single-family homes. For settlements with more than 25 persons, 706 larger infiltration facilities have been constructed as of 1983.[11,12] The idea of constructing permeable pavements as so-called Unit Superstructure has been put into practice in Sweden and during the last ten years the number of residential areas where porous pavements have been constructed have grown quickly. Porous pavement construction has a great potential for reducing and attenuating stormwater runoff. It is highly tempting to use this construction on a larger scale. Field and laboratory tests show that porous pavements have the ability to reduce pollution in stormwater on the temporal scale comparable with the length of operation of existing surfaces, i.e., approximately 20 years.[13,14] However, the portion of pollution migrating to the soil and groundwater is not really known.[15,16]

Policies of local disposal of stormwater have contributed to some improvement of the environment. However, soon it was realized that the improvement is only local and short-lasting. Ecological degradation is continuing in areas situated further apart from pollution sources. This means that problems have been only moved away from cities to previously untouched areas. In order to cope with these problems, a completely different approach is now needed.

During the last two years, the number of applications of a new approach to stormwater problems has increased quickly in Sweden. There is now a clear tendency to increase the use of surface water bodies, wetlands, and plant filters in order to reduce stormwater pollution. This requires that stormwater remains on the surface instead of flowing in conduits. By disconnecting stormwater from overloaded combined systems two very important targets may be achieved simultaneously: firstly, it may satisfy the increasing environmental concern of the public, and secondly, it may be an economic solution. A major part of combined and separated sewer systems in Sweden is rapidly aging and in urgent need of renovation and/or replacement. These very costly measures can be decreased if stormwater volume is reduced by surface disposal.

In the city of Malmö, several projects involving the natural treatment of stormwater are under construction. In the so-called Toftanäs Project, storm water from 200 ha of newly developed industrial area is routed to a 10,000-ha green area of artificially created wetland pond with a carefully designed ecological system. Total volume of the pond is 58,000 m^3.[17] Because the

Figure 1.   Toftanäs wetland and stormwater detention pond (after Movium[17]).

surrounding region is flat, the level of the whole wetland area is placed 3 m below the normal ground level in order to assure gravitational water flow through the wetland. During dry periods and small rainfall events, stormwater from several outfalls flow through a meandering stream and exits to the downstream receiving water body, the Risebergabäcken River (Figure 1). During larger rain events, the whole wetland area is flooded. Three circular higher parts within the area create islands during the rain. The plant population of different types of grass, bushes, and trees is chosen especially in order to perform treatment. The islands and edges of the area are covered by Salix. In the anaerobic conditions present in the Salix rootzone, heavy metals adsorb to soil particles. The aerobic zone of flooded surface and stream provides very effective treatment, especially from nitrogen and phosphorous. In spite of the unfavorable topographical situation of the area which required expensive excavations, the total cost of the project was about four million Swedish crowns less than the traditional solution utilizing stormwater conduits.

Another example of activities going on in Malmö, is the so-called "Stadium Project".[18] The combined sewer system in the southwestern part of Malmö is often overloaded. Several overflow structures deliver approximately 190,000 m^3 water to the receiving channel. Stormwater from a 350 ha area will be

**Table 1. Change in Pollution Load in Receiving Water after
Implementation of the "Stadium Project"**

	Change of Pollution Load, (tons/year)	
	CSO + Treated Wastewater	Combined Effect, Release to Receiving Water
Nitrogen	– 13.0	– 11.98
Phosphorous	– 0.35	– 0.17
Lead	– 0.004	+ 0.0046
Zinc	– 0.05	+ 0.06
Copper	– 0.07	– 0.04
Suspended solids	– 18.10	+ 41.40
Bacteria	– 20×10^9	– 14×10^9

disconnected from the combined system and will flow on the surface in an artificially-created stream through the park area to the receiving channel. Wetlands and periodically flooded zones along the new water course will create an attractive park, increasing the aesthetic value of the district and, simultaneously, treating stormwater. The wetland area is divided into three zones with different designs and functions. The first part, *the mouth,* is thought of as a transition zone between the urban area and the green area of the park. Design of the otherwise trivial outlet of the large conduit was performed with the goal of achieving the look of a natural water source with a creek winding through stones and green meadows. The second part, *the jungle* will have a dense and differentiated plant cover. Pollution levels in water seeping through this part of wetland will be significantly reduced. In the third section, the *meadow land*, the stream meanders through an open meadow landscape which can be used for possible recreational purposes during dry periods. Realization of the project will bring several benefits compared with the present situation: (1) the volume of CSO will decrease by approximately 25,000 m³/year, (2) the volume of water coming into the treatment plant will be reduced by 675,000 m³/year, (3) the overloading of the combined conduit will decrease and reduce flooding problems, and (4) the aesthetic and recreational value of the region will increase. Environmental benefits are not so obvious because presently most of the stormwater volume goes through the treatment plant. A very conservative calculation of pollution budget for present and future situations is presented in Table 1.

Increases in the release of lead, zinc, and suspended solids due to the increased stormwater volume will probably be much less because uptake of these substances in the wetland must occur. This uptake was not included in the calculations above because the exact reductions are unknown. The performance of the construction will be closely followed and real benefits/drawbacks will be evaluated.

2.4 APPLICATION EXAMPLE: POLAND

2.4.1 Background

The City of Zabrze, situated in the Upper Silesian Region of Poland, has been classified as one of the most degraded places in the world. With about 198,000 residents in an area of 80 km_2, plus three coal mines, open-hearth furnaces, cokeries, chemical factories, five thermal electric power stations, and other industrial facilities situated within the densely populated area, it provides a classical example of environmental degradation at every level.[1,16]

Water resources are scarce in this region, and parts of the city are periodically excluded from water supply. Heavy industry consumes large quantities of water and releases polluted effluents without any treatment. Most municipal waste waters are discharged to the rivers without treatment or after mechanical treatment only. Eighty percent of the yearly flow of the River Bytomka, which receives all effluents, consists of industrial and municipal waste waters.[2,19] About 70% of all households in the town are heated by coal stoves with the remaining 30% connected to the central heating plant. Thermal plant and house stoves burn coal with 1.3% sulfur content (Swedish plants use coal with 0.3% sulfur). Neither the existing nor the newly-built thermal plants are equipped for any treatment of combustion gases. The new thermal plant requires cooling water of higher quality compared with the old one which uses wastewater from River Bytomka. The only available solution for the moment is to use communal drinking water. This will reduce the existing drinking water supply by 40,000 m^3 daily.

In this situation, only a multiobjective approach, involving all urban activities within the area and providing a complex solution, may have a chance to brake the vicious circle presented above. This is the aim of newly started Polish-Swedish multiobjective restoration project.

2.4.2 Multiobjective Restoration Project

The Polish-Swedish cooperation project, partially financed by the Swedish Government, was launched in order to perform complex environmental restoration of a Mikulczycki brook catchment situated in Zabrze, Poland. The overall goal of the presented project is to transfer modern, ecologically sound, and economically efficient approaches and technologies dealing with environmental protection from Sweden to Poland. The project aims to show that economically weak countries should implement modern ways of thinking and apply integrated solutions, since these solutions are often the most economically effective in the long run. The project is based on an assumption that the best environmental actions which create the most ecologically sound solutions are based on pollution prevention achieved by modern small-scale technology. As an example of an application of this assumption, let us look at the detailed goals and general outline of the proposed solution. The detailed goals of the project may be divided into four parts:

1. Reduce pollution of the Mikulczycki brook to the degree necessary for industrial usage of the water. This would provide about 40,000 m^3 of water necessary for the cooling of the new thermal plant. After the introduction of the new thermal plant, new residential areas will receive central heating. This will reduce air pollution because it will no longer be necessary to burn coal in single houses and flats. The following measures are proposed:

    • The "key operation", which is necessary to reduce the pollution level in the brook to the degree that other biological measures are effective, will be achieved by constructing the main sewer collecting municipal wastewaters from the upper, small, densely populated part of the area (Helenka). Allocation of funds has been given first priority.
    • Upgrading of treatment capacity of the existing treatment plant by introducing aeration and constructing a biological pond filter with root zone uptake. The existing treatment plant was constructed during the 1930s.
    • Improving incoming water quality by protective measures in the upper reaches of the brook.
    • Revising production methods and wastewater generating processes in all industries within the area in order to reduce pollution releases, and the effluent volume at the source.
    • Disconnecting the most polluting industries from the river and sewage system. Local treatment of effluents from the three most polluting industries using biologically balanced ground infiltration with pre-aeration.
    • Constructing small, local treatment units in order to selectively treat the effluents from the most polluting tributaries.
    • Increasing the residence time of water and the self-purification capacity in the river by 50% through the changing of straight reaches to meandering courses.
    • Creating vegetative filter strips and seasonally-flooded riparian zones along the course of the brook and its tributaries.
    • Creating small lakes designed as stormwater detention basins along the course of the brook.
    • Reducing the volume of wastewater from one residential area by about 30% using water saving schemes and techniques.

2. Reduce water consumption in one housing area of the catchment using water saving technology. This will reduce wastewater volumes and

drinking water shortage. About 30% reduction will be achieved by the installation of water saving toilets, the use of more efficient cocks and better packings, the restoration of water networks and the installation of water meters in each lodging.

3. Reduce energy consumption in the Helenka housing area by decreasing heat loses. Since 70% of houses are heated by coal stoves burning coal, the energy savings become very important for the protection of the environment. Energy-saving measures include:

    - Installation of regulators in the heating network.
    - Installation of thermostats and instructing residents.
    - Introduction of electricity saving schemes and energy-efficient incandescent lamps.

4. Upgrade the aesthetic appearance of the Helenka housing area. The goal is to create a positive example of a case site where environmentally and economically sound management may also increase aesthetic values of the area; thereby, indirectly increasing the quality of life for its inhabitants. The program includes upgrading green areas, improving pavement standards, painting houses, and introducing better routines for disposal and removal of solid wastes.

Since the Polish-Swedish restoration project contains elements from several disciplines (physical planning, water resources planning, wastewater treatment by conventional and biological methods, management of energy, design of aesthetical aspects, etc.), the multidisciplinary team of researchers which has been created consists of hydrologists, city planners, biologists, ecologists, economists, and politicians. Polish and Swedish sides are equally represented during the working sessions of the team. It is noteworthy that all elements of the project solution are mutually interdependent, therefore, all elements must be performed in a logical sequence. For example, all biological measures in the river would not function without the disconnection of the majority of municipal and industrial waste discharges. Stormwater flow must be attenuated before plant filters may be effective. Public acceptance and participation is essential for the success of the project.

2.5 SUMMARY AND CONCLUSION

Our perception of the world and its problems has recently changed. We now realize the global nature of environmental problems, and we know that environmental degradation is continuing in developing countries and in Eastern Europe.

Cities of industrialized countries still contribute, in spite of all the efforts and costs, to the continuing degradation of the global environment. The advanced sewerage systems and sophisticated treatment plants now in use do not eliminate the most hazardous substances. Stormwater delivers heavy metals and other nondegradable toxic elements to the environment.

The result is that, increasingly, the traditional "end of pipe" solutions are losing credibility. The goals of water management have already been reformulated from traditional storm drainage to more ecologically sound solutions. The new goal is to avoid accumulation of nondegradable hazardous substances and to adapt the waste output to what nature can endure.

Expansion of large cities is still continuing at an accelerating pace. The world's fastest growing and largest cities are situated in the regions with lowest gross national product (GNP). To say that these cities create environmental problems is an understatement. Just to keep pace with the population increase in the less developed countries, another 1.3 billion people must be supplied with water, sanitation, and waste handling. These facts raise the question to what extent our present wastewater treatment solutions are realistic alternatives for developing cities.

There is a new realization that the struggle against global pollution can only be efficient if water management is integrated with the management of other related urban subsystems such as solid wastes, industrial production systems, energy production and consumption, transportation systems, etc. This approach is based on a deeper understanding of the large, long-term circulation systems of global scale that involve all energy and materials. It is evident that development of new, promising technologies requires a merging of knowledge from several scientific disciplines. This is difficult because it requires an effort to leave traditional ways of professional thinking and to create links to other related disciplines. Biology, ecology, and ecological engineering are examples of sciences that can contribute much to the development of new thinking and new technical solutions. Unfortunately, there is practically no communication or exchange of viewpoints between sanitary engineers dealing with traditional treatment and ecologists developing new biological systems.

Some basic conditions must be fulfilled before these new technologies can be applied more widely. First, new approaches and new technologies must be integrated with existing traditional technology and infrastructure, instead of fighting against it or trying to replace it. Second, the management of wastewater must be integrated with the management of all other types of water and wastes within the catchment and the management of all human activities such as industrial production, energy production and consumption, transportation, etc.

Performing small, well-defined pilot projects where the transfer of ideas and approaches is the major goal (instead of the construction of facilities designed according to a previously governing technical paradigm) is perhaps the most important action to which financial means should be allocated.

Stormwater pollution constitutes only a small part of total pollution load, but still it must be considered together with all other pollution sources.

REFERENCES

1. Konstantynowicz, E., Ed. "Problems of protecting environment and natural resources in Katowice region," PTPNOZ, Sosnowiec (September 1986), in Polish.
2. Postawa, K., J. Kapala and M. Kostecki. "Concept of restoration of water resources and environment in Bytomka river catchment," Polska Akademia Nauk, Instytut Podstaw Inzynierii Srodowiska (Zabrze 1986), in Polish.
3. Ontario Ministries of Natural Resources, "Urban Drainage Design Guidelines," Environment, Municipal Affairs and Transportation and Communication, Association of Conservation Authorities of Ontario, Municipal Engineering Association, Urban Development Institute, Ontario, (February 1987).
4. Sundblad, K. "Recycling of wastewater nutrients in a wetland filter," Ph.D. Thesis, Linköping Studies in Arts and Sciences (1988).
5. Reddy, K.R., and T.A. DeBusk. "State of the art utilization of aquatic plants in water pollution control." *Water Sci. Technol.* 19(10): 61 (1987).
6. Ridderstolpe, P., and I. Kindvall. "Aquaculture, water treatment and reuse of resources," (in Swedish), Royal Institute of Technology, Deptartment of Land and Water Resources, Meddelande Trita-Kut 1050, Stockholm (1989).
7. Zweig, R.O. "Freshwater aquaculture in China — ecosystem management for survival," *Ambio*, 14(2): 66 (1985).
8. Stahre, P. "Flow attenuation in sewerage systems," Byggforskningsrådet T13 (1981), in Swedish.
9. Falk, J. "Urban Hydrology," Report No. 3045, Department of Water Resources Engineering, University of Lund, Sweden (1980).
10. Niemczynowicz, J. "Outcomes of Possible Water Management Policies on Water Quality from the City of Lund, Sweden," *Proceedings of the International Symposium on Urban Hydrology and Municipal Engineering*, (Toronto, Canada: Markham, 1988), p. 8.
11. SNV. "Infiltration of wastewater: background, function and environmental impacts," Nordisk Samproduktion Naturvårdsverket, Nordiska Ministerrådet, (1985), in Swedish.
12. SNV. "Infiltration of municipal waste water), Statens naturvårdsverk, Liber förlag, Buhuslänningens Boktryckeri AB Uddevalla (1985a), in Swedish.
13. Malmquist, P.A., and S. Hard. "Groundwater Quality Changes Caused by Stormwater Infiltration," *Proceeds of the Second International Conference on Urban Storm Drainage* (Champaign-Urbana, 1982), p. 89.
14. Niemczynowicz, J., W. Hogland and T. Wahlman. "Consequence Analysis of the Unit Superstructure," *Vatten.* 41(4.85): 250 (1985), in Swedish.

15. Niemczynowicz, J. "Permeable Pavements - Solution of all Problems or a Threat for the Future?," *Proceeds of the International Conference Topical Problems of Urban Drainage.* (Czechoslovakia: Strbske Pleso, April 1988).

16. Niemczynowicz J. "Report from an ecological catastrophe," Miljödelegationen Västra Skåne, Info 3 (October 1989), in Swedish.

17. Movium, VAV. "Place for rainfall," Stad och Land No 86, Hässleholm (1990), in Swedish.

18. Larsson, T., and A.C. Sundahl. "Stationparken — ett projekt om öppen dagvattenavledning," Ph.D. Thesis, Department of Water Resources Engineering, University of Lund (1990), in Swedish.

19. OBKS, "Prognosis of water quality in Mikulczycki brook," BKS 05/93/83/BWI (1988), in Polish.

3 THE INTEGRATED APPROACH TO CATCHMENT-WIDE POLLUTION MANAGEMENT

3.1 INTRODUCTION

During the last century civil engineers provided the urban drainage infrastructure which assisted in curing the public health problems in the U.K. resulting from the development and urbanization associated with the Industrial Revolution. This knowledge was then used to build similar systems in towns and cities throughout the world. Today engineers and scientists are tackling the next generation of problems — the impact of urbanization upon the wider environment.

The creation of the U.K. Water Industry into large water companies with environmental regulation by the National Rivers Authority has resulted in the development of approaches to pollution management which are cost effective. Consideration of complete river basins has focused on the need to adopt an integrated approach to ensure maximum benefit is achieved for both the utility and the regulatory authority.

The U.K. has pioneering techniques for sewerage rehabilitation and implemented these to a greater extent than any other country. WRc (the Water Research Centre) was instrumental in developing and introducing these techniques and has been advising on the potential benefits of applying them around the world.

3.2 WHAT IS WRc?

WRc is an independent private company with a staff of 650 and turnover in excess of $60 million per annum. It operates from two principal sites: the Headquarters and Environmental laboratory situated at Medmenham, some 50

0-87371-805-4/93/$0.00+$.50
© 1993 by Lewis Publishers, Inc.

km west of London, and the Engineering laboratory at Swindon a further 100 km to the west. It also has offices in Scotland, Philadelphia (USA), Hong Kong, and Bologna (Italy).

Although WRc's background is in research, specializing in the strategic, planning, and management aspects of engineering and environmental problems, it places great emphasis on the implementation of its research results and in developing practical and cost-effective solutions to difficult problems.

Its expertise covers the whole water cycle from water resources, through drinking water supply and wastewater collection, to effluent and sludge disposal. Outside the water cycle, WRc also carries out general environmental research on subjects which include aspects of industrial water management, waste disposal, land contamination, and air pollution. It has developed sophisticated models for many purposes, particularly for marine pollution.

As a private company, WRc has extensive links with the European Commission, and an important part of its work is to consider the strategic implications for its clients of trends and developments in environmental policy and legislation. Its strength is in its ability to examine problems in a strategic context and to draw on the results of its research to find integrated and cost-effective procedures for dealing with environmental and engineering problems.

3.3 APPROACHES TO POLLUTION CONTROL

Water quality surveys enable rivers and estuaries to be classed, depending on the ability to support fish life and the water's suitability for such things as treatment for potable use, agricultural use, or as a bathing water in the marine environment.

Within the general acceptance of the desirability of protecting the aquatic environment there is agreement that control should be by consenting discharges to the sewerage, river, or marine system. Differences of opinion come in the approach to setting and monitoring of consents.

There are two possible alternative approaches to pollution control: by specifying either the treatment technology to be used or the level of allowable receiving water impact that will be tolerated. The former is a Best Available Technology/Uniform Emission Standard (BAT/UES) approach favored in many parts of Europe, while the latter is the Environmental Quality Objective/ Environmental Quality Standard (EQO/EQS) approach traditionally applied within the U.K.

An argument in support of the BAT/UES approach is that it is equitable to specify a uniform emission standard with which all discharges have to comply, or the type of treatment plant that may be installed, via the best available technology. However this approach can be expensive and is difficult to apply to an existing system requiring only minimal upgrading. Frequently it may be ineffective because it can only be applied to the point of source loadings.

Enforcing the best available technology (BAT) may show some improvement in receiving-water quality initially, but as the nonpoint source pollutant loads build up, a stage is reached where a total catchment approach will be necessary. This is the situation in the U.K. and thus applying the EQO/EQS approach is the most cost-effective and currently preferred approach. However there is a danger that advantage is not taken of readily available pollution reduction options because the quality standard is easily achieved.

Environmental quality objectives (EQOs) and associated standards (EQSs) are defined for the receiving-water bodies. Models are then used to quantify the environmental impact of existing and any proposed discharges, and their interaction with each other, so that discharge consents can be agreed that will not put the water quality standards at risk. The EQO/EQS approach is much more demanding in investigation and administration costs, but all parties can have confidence that the solution agreed upon will work and that the cost savings from the generally cheaper capital works will more than justify the extra effort.

For the purposes of the management of urban drainage, the environmental problems to be solved may be summarized as follows:

* recognition of the pollutant load and concentration increase in receiving waters from urban discharges such as from storm overflows
* quantification of the impact of these increases on aquatic life and on the suitability of the water for its intended uses
* development of standards (EQSs) to enable the discharges to be controlled so that the desired uses are not put at risk
* development of a river water quality monitoring/reporting system that can handle the short-term nature of urban drainage discharges

3.4 TOWARDS INTEGRATED POLLUTION MANAGEMENT

The bulk of a water utility's assets are the underground pipes, whether sewers or water mains, and upgrading the performance of these accounts for a major proportion of revenue and capital expenditure. Cost-effective pollution management requires a sound knowledge of the system performance, and pollution management studies should be built into asset management plans.

Upgrading the performance of the sewerage systems through the undertaking of drainage area studies using the Sewerage Rehabilitation Manual (SRM)[1] procedures is one component of urban pollution management. Technology is available to assess the condition of sewers and quantify the hydraulic performance, which leads to improvements in the system's performance being implemented. The concept of undertaking drainage area studies using the SRM procedures in order to produce planned sewerage rehabilitation programs is now well established in the U.K. The comprehensive system and performance data once obtained can have many uses:

* effective planning of the capital works program
* quantification of future funding needs
* demonstration of standards of service being achieved

To obtain the same benefits within the environmental area, it is necessary to produce a complementary catchment plan (CMP) built around hydrodynamic and water quality models of the receiving waters.

3.5 POLLUTION MANAGEMENT PLANNING

In order to apply the Environmental Quality Objective/Environmental Quality Standards (EQO/EQS) approach to urban pollution control, collaboration is necessary between those discharging to the environment and the organizations charged with regulation. To achieve the necessary environmental benefits, it is to the benefit of all the organizations concerned to collaborate, whether it be the sewerage operators, industrial dischargers, farmers, or regulators. The public's concern with the environment should ensure that the regulators protect the aquatic environment and that the water utilities and dischargers manage their businesses efficiently within the financial and regulatory framework.

A suitable framework for developing a Catchment Management Plan is shown in Figure 1, and is being developed as part of the Urban Pollution Management research program, a collaborative project of the U.K. Water Utilities, National Rivers Authority, Department of the Environment, and the Science and Education Research Council.

Its implementation for each catchment can be considered in five phases:

Phase 1: Environmental Assessment
Phase 2: Modeling and Impact Assessment
Phase 3: Engineering Options
Phase 4: Catchment Management Plan
Phase 5: Implementation/Monitoring of Plan

3.5.1 Phase 1

The region is divided into catchments and the uses are defined for the receiving waters. Appropriate standard EQOs and EQSs are defined for these uses and the current performance monitored against these standards to identify the priorities for pollution abatement. Liaison is necessary with those dischargers within the catchment to assess how far their actions are responsible for any perceived problems and to review their in-house data collection and modeling abilities.

3.5.2 Phase 2

In the second phase, flow and water quality computer models are built for the urban drainage, river, and marine systems. The urban drainage system will discharge pollutant loads into the river. These will be both continuous, from

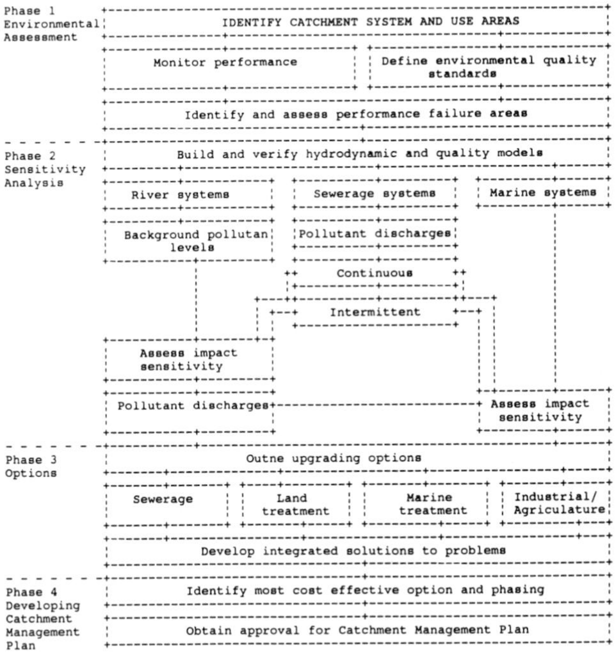

```
Phase 1      +-------------------------------------------------------------+
Environmental|           IDENTIFY CATCHMENT SYSTEM AND USE AREAS           |
Assessment   +--------------------------------+----------------------------+
             +-----------------+--------------+ +------------+--------------+
             |    Monitor performance         | | Define environmental quality|
             |                                | |          standards         |
             +-----------------+--------------+ +------------+--------------+
             +-------------------------------------------------------------+
             |        Identify and assess performance failure areas        |
- - - - - - -+----------------------------------+--------------------------+
Phase 2      |      Build and verify hydrodynamic and quality models        |
Sensitivity  +----------+------------+---------+----------+---------+------+
Analysis     +----------+------------+ +-------+----------+ +-------+------+
             |   River systems       | |   Sewerage systems | | Marine systems |
             +----------+------------+ +-------+----------+ +-------+------+
             +----------+------------+ +-------+----------+
             | Background pollutant  | |Pollutant discharges|
             |        levels         | +--------+---------+
             +----------+------------+     ++         Continuous        ++
                        |             +----+----------------------+---+|
                        |             |  +--+  Intermittent  +--+  |  |
                        |             |  +----------------------+  |  |
             +----------+-------+---+ |                            |  |
             |  Assess impact     | |                             |  |
             |  sensitivity       | |                             |  |
             +----------+---------+ |                             |  |
             +----------+---------+ +----------------------+ +-+--+--+------+
             | Pollutant discharges+---------------------------+ Assess impact |
             |                    |                        | |   sensitivity  |
             +----------+---------+                        +--------+-------+
- - - - - - -+----------+--------------------------------------------+------+
Phase 3      |              Outne upgrading options                         |
Options      +--------+---------------+------------+----------+--------+----+
             +--------+------+ +------+-------+ +--+-------+ +--------+----+
             |   Sewerage    | |    Land      | |  Marine  | | Industrial/  |
             |               | |  treatment   | | treatment| | Agriculture  |
             +--------+------+ +------+-------+ +--+-------+ +--------+----+
             +--------+--------------------------------------------+------+
             |        Develop integrated solutions to problems            |
             +-------------------------------------------------------------+
- - - - - - -+-------------------------------------------------------------+
Phase 4      |        Identify most cost effective option and phasing       |
Developing   +-------------------------------------------------------------+
Catchment    +-------------------------------------------------------------+
Management   |        Obtain approval for Catchment Management Plan         |
Plan         +-------------------------------------------------------------+
```

Figure 1.   Catchment management plan flow diagram.

treatment works, and intermittent discharges of stormwater and sewage. The river model is used to check on the impact of the urban drainage and agricultural discharges and to quantify the pollutant loads passing out to sea.

The marine model can then be used to review the impact on marine uses, such as bathing and shell fisheries, of the river flows and direct discharges to the estuaries and sea. Other inputs to be integrated into the Catchment Plan are the policies of sludge disposal, land use, and the incorporation of protection zones.

The models identify the causes and scale of failures and these will need to be compatible, as the output from one model will be the input for another. With regard to the sewer, river flow, and water quality models in the U.K., this is being achieved by nationally agreed procedures.

At the end of Phase 2, an indication is provided of the level of change to pollutant loads necessary to achieve the environmental quality standards.

3.5.3 Phase 3

The dischargers, at this stage, have the opportunity to use their engineering skills to develop cost effective outline designs which can satisfy the EQS requirements. The water utilities will build these into forward planning sewerage and sewage treatment capital works programs and sludge disposal strategies as with the other discharges for industrial and farm waste treatment and disposal. This phase, because of the many interactions, will have a number of iterations and require careful coordination.

3.5.4 Phase 4

At this stage discharge consents can be set to enable the water utilities and other organizations to complete the capital program and operational procedures. The plan also sets out the timescale for environmental improvements and quantifies the resulting benefits.

3.5.5 Phase 5

In the final phase, the dischargers will implement their construction and operational programs, while the regulators can monitor the programs, consent compliance, and the resulting effects. Following the issue of new directives, it will be necessary to update the plan, which should in any event be undertaken at agreed intervals.

3.6 SEWERAGE ENGINEERING

Intermittent, storm-related discharges are poorly understood in terms of their elements, effect, and control. Yet to achieve EQOs, they must be managed effectively and be compatible with the continuous discharges. An understanding of the environmental impact of intermittent discharges is necessary for control to be achieved. Effects, such as short-term changes in concentration, coupled with duration of exposure and return period, are needed to produce the environmental quality standards for the receiving waters. Compliance with the set standards will of necessity never be fully monitored in the field, but the standards used as design criteria and control levels established through use of the models.

3.6.1 Modeling Tools

Water quality models are required for both the sewer system to predict the pollutant loads being discharged, and for the receiving waters to quantify the environmental impacts. An essential requirement of these models is that they can predict the variation in pollutant concentration in the flow during the storm. In addition, models are required to predict the performance of treatment works during and following storm loading so that the impact of time varying effluent discharges can be included in the receiving water models.

3.7 CONCLUSIONS

Many two- and three-dimensional hydrodynamic/dispersion models are available for simulating the impact of discharges from long sea outfalls, but in many marine areas, the cause of failure is at least in part due to bacterial loading from the intermittent discharges from combined sewer overflows and surface water outfalls. Application of these models in conjunction with sewer flow and quality models and rainfall time-series demonstrates the effectiveness of proposed coastal sewerage upgrading works.

A similar approach has been adopted for river pollution, where again there are continuous and intermittent discharges. In this case the emphasis has been on developing a river impact model which can simulate the chemical and biological effects of the shock loadings.

Effective pollution control requires a detailed catchment-wide appreciation of the impact of current pollution loadings on the total receiving-water systems and the implications for the associated sewerage systems of any proposed modifications to the discharge consents.

The integrated approach described provides greater confidence that environmental quality standards will be achieved than rigid adherence to uniform emission standards or best available technology, and will show substantial savings in capital cost of the upgrading works.

REFERENCES

1. WRc/WAA. *Sewerage Rehabilitation Manual*, 2nd ed. (1986).
2. Clifforde, I.T., A.J. Saul and J.M. Tyson. "Urban Pollution of Rivers — The U.K. Water Industry Research Programme," Proceedings of the International Conference on Water Quality Modelling in the Inland Natural Environment, Bournemouth, BHRA (June 1986), pp. 485-491.
3. Fiddes, D., and I. T. Lifforde. "River Basin Management — Developing the Tools," IWEM Conference: Technology Transfer in Water and Environmental Management, Birmingham, AL (September 1989).
4. Powlesland, C. "A Computer Simulation Model (WISDOM) for Identifying Cost-Effective Sludge Treatment and Disposal Options," WRc PRU 1466M (1987).
5. Powlesland, C. "The Development of a Sludge Disposal Strategy," WRc Conference Proceedings on Alternative Uses of Sewage Sludge (1987).
6. Fiddes, D. "The U.K. Approach to Catchment-Wide Pollution Management," in Proceedings of the International Symposium on Integrated Approaches to Water Pollution Problems, SISIPPIA '89, Vol.2 (1989), pp. 169–177.
7. Ellis, J.C. WRc Manual on the Design and Interpretation of Monitoring Programs. (1989).
8. Rumsey P.B., and T.K. Harris. "Asset Management Planning and the Estimation of Investment Needs," Urban Water Infrastructure NATO ASI Series (1990).

4 CATCHMENT MANAGEMENT PLANNING: U.K. EXPERIENCE

4.1 INTRODUCTION

4.1.1 The National Rivers Authority (NRA)

The NRA was formally established under the 1989 Water Act on September 1, 1989, and has statutory duties and powers in relation to water resources, pollution control, flood defense, fisheries, recreation, and conservation and navigation in England and Wales. The Authority has a head office and ten operational regions based on the surface water catchments of the most significant river basins in the two countries. In 1989/90 the Authority had a budget of £335 million and employed over 6500 staff. Over 70% of the NRA's income comes from charging schemes and regional government, the remainder from central government.

In managing its activities, the Authority has adopted the following mission statement to guide its role as "guardian of the water environment".

The NRA will protect and improve the water environment. This will be achieved through effective management of water resources and by substantial reductions in pollution. The Authority aims to provide effective defense for people and property against flooding from rivers and the sea. In discharging its duties, it will operate openly and balance the interests of all who benefit from and use rivers, groundwaters, estuaries and coastal waters. The Authority will be businesslike, efficient and caring towards its employees.

Thames Region of the NRA encompasses the majority of Greater London and drains an area of over 12,900 km². The population of the region is over 11.6 million. Thames Region has the largest number of staff among the ten NRA regions employing over 1300 staff.

0-87371-805-4/93/$0.00+$.50
© 1993 by Lewis Publishers, Inc.

4.1.2 The Catchment Management Planning Process and the NRA

At the strategic level, catchment management planning can be considered to be a process which balances society's need for development with the need to conserve and enhance the river environment. In operational terms, it is a means of ensuring that the functionally organized activities (e.g., flood defense, water resources, water quality) of the NRA are mutually supportive within a catchment and compatible with the Authority's mission statement.

The output from the process is a framework for managing a complete river system that involves the whole range of individuals and groups (statutory and nonstatutory) interested and involved in the river catchment. Although relatively new in the United Kingdom, similar principles have been applied in a number of countries.[1]

Ideally, a catchment management plan (CMP) should encompass all the functions of the NRA as well as all the other influences (e.g., land use planning and agricultural practices) on the river environment (see Figure 1).

It should detail the NRA's vision for the river catchment and the actions that the NRA and other interested parties should fulfill in order to achieve that vision. To be successful it must, therefore, be based both on an understanding of how the river system functions and is influenced and on wide consultation with all those who will play a part in its achievement (see Figure 2). A primary objective of the CMP is to ensure that the catchment is managed in an integrated and proactive way rather than in a single functional and reactive way.

4.2 CATCHMENT MANAGEMENT PLANNING IN NRA THAMES REGION

The development of catchment management planning for rivers in the Thames basin has been in response to a number of influences. These include growing national awareness and concern over continued exploitation of natural resources, heavy pressure on the land use planning system in the region leading to uncontrolled direct and indirect damage to the river environment, and growing staff awareness of the potential synergy to be gained from focusing NRA functions and external agencies on the key issues in a particular river catchment.[2]

In early 1989 the NRA Thames Region published its Implementation Guidelines detailing the methodology for the preparation of CMPs (see Figure 3).[3] This led to the production of Evaluation Reports for three test catchments by the end of the same year. The aim of these reports is to present, in qualitative and quantitative terms, what information exists on the catchment and what investigations are required to produce the CMP.[4-6] An important part of this process is the identification of the perceived key issues affecting the river environment in the catchment. Completion of the studies identified in the Evaluation Report and their subsequent synthesis results in the continuation of key issues and the production of the CMP itself. The remainder of the paper describes the production of the CMP for the River Stort which was completed in early 1991.

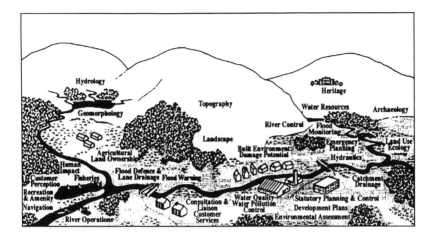

Figure 1.   Catchment management planning considerations for a sample river catchment.

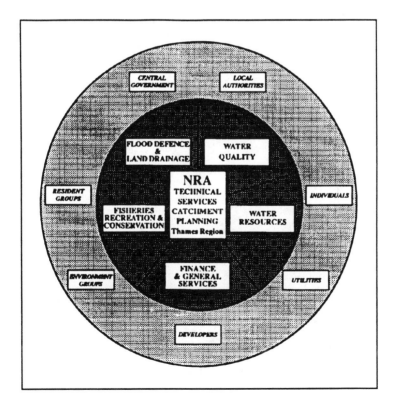

Figure 2.   NRA organized activities and interested parties in the catchment management planning process.

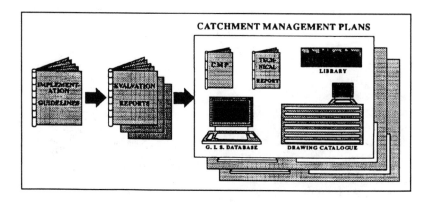

Figure 3.   Catchment management plan preparation methodology.

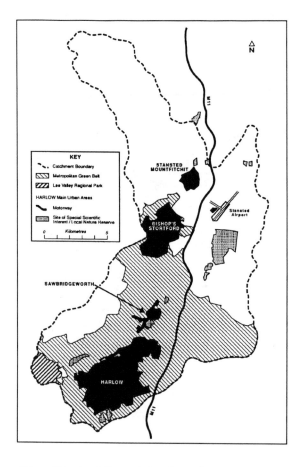

Figure 4.   River Stort catchment: land-use features.

4.3 RIVER STORT CATCHMENT

The River Stort catchment is located to the northeast of Greater London (see Figure 4) and covers an area of 280 km^2. Although the centers of development, including Harlow and Bishop's Stortford, have a total population of about 130,000, the most significant land use is agriculture, to which 85% of the catchment area is devoted. Bisecting the catchment are regionally important rail and motorway (M11) links between London and Cambridge. In the north of the catchment, London's third airport, Stansted, has just opened. This airport has a capacity of eight million passengers per annum.

An older communication link is the Stort Navigation, which runs south from Bishop's Stortford to the River Lee, predominantly along the line of the River Stort, which forms part of the administrative boundary between the counties of Hertfordshire and Essex. This link was opened in 1769 to transport malted barley (for the making of beer) to central London. The navigation is no longer used for commercial traffic but Londoners still drink beer!

The ecological habitat and landscape of the Stort Valley, almost unique in southern England, has been partly created as a result of the construction of the navigation. Its value has been threatened on many occasions by activities, such as agricultural drainage and insensitive flood defense works but remains essentially intact.

4.4 CATCHMENT ISSUES

The Evaluation Report for the Stort catchment identified the following key issues as a result of a desk top study and discussions with NRA Thames Region staff and external parties:[5]

- standards of urban flood protection in the Stort valley and the need for flood defense works
- a joint river control strategy for the Stort Navigation with the navigation authority (British Waterways)
- the impact on NRA interests of development pressures associated with Stansted Airport and the M11 corridor
- the water-dependent habitats in the catchment require recording, protecting and enhancing
- NRA interests, including surface water runoff management, preservation of flood plains, and river corridor continuity should be catered for in the statutory land-use plans for the area

4.5 BASELINE SURVEYS

In order to evaluate the above key issues it was necessary to complement existing knowledge of the catchment with a series of baseline studies.

These included:

- hydrological modeling of the catchment using RORB[7]
- computational hydraulic modeling of the channel and floodplain system in the Stort Valley using ONDA[8]
- appraisal of the natural resources of the river corridors, eg., ecological habitat, landscape, archaeology, recreation and amenity, river morphology
- land use appraisal and identification of future pressures through a review of national, regional and local planning policy, guidance and data

Water resource and quality issues could not form an integral part of the overall study due to other priorities within Thames Region. This weakened the integrity of the process outlined in Sections 1 and 2 but did not prevent the investigations from progressing successfully in respect of the NRA's flood defense, fishery, conservation and recreation functions, and liaison with external interest groups.

The baseline surveys brought together a range of internal expertise and external consultants into a very focused project team which needed to undertake not only technically diverse studies but also an extensive consultation and liaison program. This part of the study brought in statutory authorities involved in land use planning (the County and District Councils), sewerage and sewage disposal (the Water Utility Public Limited companies and their agents), nature conservation (the Nature Conservancy Council), recreation (Sports Council), landscape (Countryside Commission) and navigation (British Waterways), among others. It also involved nonstatutory groups, particularly in nature conservation (Herts and Essex Wildlife Trusts), and individuals with particular knowledge of the catchment.

Many of the above statutory authorities have common responsibilities within the catchment because the river forms a significant administrative boundary. This has meant that no single body has ever considered many of the needs of the natural river system and its catchment. Consequently different initiatives and procedures have been implemented on either bank of the same river system. The statutory land-use plans, for example, when viewed in terms of the Stort catchment rather than land-use planning districts, yield significant and potentially damaging contradictions. The NRA acting as the "guardian of the water environment" is ideally placed to identify and rectify such problems through dialogue with the appropriate authorities.

4.6 ISSUES AND ACTION PLANS

One of the aims of the Stort CMP is to translate the national objectives and policies of the NRA into catchment specific actions. The process used to identify the catchment issues and the actions necessary to resolve them involved "interaction sessions". These sessions enabled those involved in the baseline survey phase to highlight the results of their studies and begin to relate them to the other studies. Through this system of meetings, it was possible to reassess the catchment issues highlighted in the Evaluation Reports and the implications of actions to resolve them on other elements of the river environment. This process is best considered by reference to three examples of issues in the Stort catchment.

4.6.1 Increased Boat Moorings on the Navigation

Issue

The navigation authority, British Waterways, is being encouraged by central government to maximize its revenue. On the Stort Navigation this need was translated into a "business plan" objective to double the number of boat moorings on the navigation. Studies undertaken for the CMP highlighted two key consequences of this action. Firstly, increasing mooring numbers would be likely to damage and degrade the sensitive ecological habitats of the navigation both directly (through increased boat movement) and indirectly (through the need for more frequent dredging). Secondly, because lock usage would increase in summer periods, the distribution of water in the system would be altered leading to an increased risk of algal blooms. This would mean that other users (e.g., canoeists) would be put at a health risk.

Action

The CMP studies enabled the consequences and scale of the British Water-ways proposal to be assessed in terms of its impact on the wider river environment. Using this evidence it was possible to persuade British Waterway to modify their proposals so that a balance between conserving the ecology and existing standards of users, and enhancing the recreational use could be achieved.

4.6.2 Surface Water Runoff Management

Issue

The River Stort catchment is under intense pressure for development in an area where the river is sensitive to increased rates of stormwater discharge. Development of the land in the catchment is controlled by five different local planning authorities, who each have a statutory duty to consult the NRA when a planning application is made. Time is of the essence, as the complete process must be completed within eight weeks if refusal and subsequent appeal procedures are not to be invoked. Within this time scale, a piecemeal control of stormwater management will result unless previous catchment wide studies have been completed.

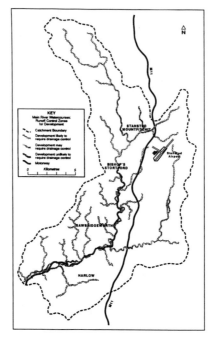

Figure 5.   River Stort catchment: main river watercourse.

Action

The extensive hydrologic and hydraulic analyses of the river catchment undertaken for the CMP has enabled an insight to be gained into the behavior of the complete drainage system which has been translated in practical terms into a surface runoff management plan (Figure 5).

The plan enables an "at a glance" decision to be made if stormwater is likely to be a problem and the planning control necessary, the test being applied to the watercourse to which any proposed development would discharge. Figure 5 clearly shows that while the main river can accept direct stormwater discharge, the need for stormwater control exists on most tributaries and increases with distance up the catchment.

In real terms, such control is not overly onerous on a development if it can be coordinated in this way. Typically the consent to develop in the areas where control is required would be to limit the storm discharge to the pre-development rates of discharge while using attenuation storage or other source control techniques to maintain the level of performance on site. On a recent project for the development of 21 homes on a 1 ha site, this involved the mobilization of some 43 m^3 of storage while limiting the outflow to 17 L/sec. The advantage of having a coordinated plan is that all parties, from developers to planners, are aware of the need for surface water runoff management right from the start of any project.

4.6.3 Spellbrook Flood Lagoon

Issue

This 70,000-m^3 flood lagoon was constructed in 1980 to compensate for lost flood plain storage in the town of Bishop's Stortford upstream. Detailed hydrological and hydraulic modeling studies for the CMP have shown that the lagoon will fail to fill other than in a catastrophic event. Ecological investigations have also shown that the floristic value of this former water meadow site is deteriorating due to the fact that it is not being flooded in winter.

Action

The NRA Thames Region, as operators of the flood lagoon, will reconstruct the inlet weir to the site in order to achieve the dual objective of sustaining the floristic species on the site and making better use of the flood attenuation capabilities of the lagoon. To improve the quality of water entering the lagoon a reed bed zone will be established immediately downstream of the inlet weir.

Several observations can be drawn from the above three examples and the issues and actions for the River Stort in general:

- The NRA has many statutory powers but will not achieve its mission statement unless it recognizes that other organizations can have a significant impact on the well-being of the water environment.
- dialogue with other organisations (statutory and nonstatutory) is a prerequisite to achieving a consensus, and therefore support, for any catchment strategy developed by the NRA.
- The implications of decisions affecting the water environment must be evaluated in overall terms rather than functional terms.
- Decisions affecting the catchment should be made in the context of an overall strategy, not local conditions.
- Technical expertise and a thorough evaluation of issues are necessary to develop a sustainable CMP.

4.7 THE STORT CATCHMENT MANAGEMENT PLAN

Guiding the identification of a strategy for the CMP for the River Stort has been the NRA's national policy framework.[9] This national framework has been given a regional context and as a result of the CMP studies a catchment perspective.[10] This linkage is illustrated by the following example which relates to issue (c) in Section 6.

- NRA NATIONAL AIMS: to provide effective defense for people and property against flooding from rivers and the sea; to conserve and enhance wildlife, landscapes and archaeological features associated with waters under NRA control.

- NRA NATIONAL OBJECTIVES: to identify opportunities for the enhancement of the environmental, recreational, and amenity facilities when undertaking flood defense works.
- REGIONAL ISSUES/TARGETS: the funding of certain conservation schemes is to be supported where flood defense work has or will be carried out.
- CATCHMENT ACTION: to rehabilitate the Spellbrook flood lagoon to provide an enhanced flood defense capability and to conserve the ecological value of the site.

The strategy developed for the Stort Catchment is based on the synthesis of all the catchment issues and actions that were identified (see Section 6) during the study.[11] Backing up this strategy are a number of baseline surveys and operational models (e.g., the ONDA hydraulic model). In order that this strategy can be implemented by Thames Region staff, every opportunity is being taken to ensure that the detailed investigations are translated into the appropriate format for day-to-day usage. This element of the CMP process is still continuing and involves a number of initiatives, including the use of a geographical information system and relational database.[12] In the future, greater emphasis will be placed on this part of the process, since it is impossible to implement the strategy without the right tools.

Further consultation on the plan is also continuing as part of the ongoing process of implementing and monitoring the CMP. These two elements of the whole process have been inadequately considered so far. Particular consideration will be given to the monitoring of the CMP strategy in order to "quality control" the whole process during its ongoing implementation.

4.8 CATCHMENT MANAGEMENT PLANS AND THE NRA: THE FUTURE

At the same time as the NRA Thames Region was progressing work on its CMP initiative, an NRA national working party was preparing guidelines for the preparation of such plans by each of its ten regions. The national working party considered the work of Thames Region in preparing its guidance and subsequently incorporated many of the principles outlined in this paper. Most importantly, the national group recognize that to be successful, a CMP should consider all the influences on the river environment and involve a dialogue with groups outside the Authority during its preparation.

4.9 SUMMARY

Guidance on the framework for catchment management planning will evolve as experience grows in the NRA. This process may also be influenced by changes in the land-use planning system and European Commission legislation on the need to apply a form of environmental assessment to strategic plans produced by public bodies.

Catchment management plans are now recognized as an essential element for the successful achievement of a better water environment in England and Wales. The challenge for the future is to make the process of catchment management planning an operational success. Experience gained in Thames Region on the River Stort CMP indicates that the process is both feasible and beneficial but that adequate consideration has to be given to the implementation and monitoring phases.

DISCLAIMER

The views expressed are not necessarily those of the National Rivers Authority.

REFERENCES

1. New Zealand Regional Catchment and River Board. "Clutha, Kawarau and Hawea River catchment management plan," (1986).
2. Gardiner, J.L. "Riverbank conservation the wider view," *River Conservation Conference*, Hatfield Polytechnic (April 1990).
3. National Rivers Authority Thames Region. "River catchment plans for flood defense and land drainage," 2nd rev. (September 1989).
4. National Rivers Authority Thames Region. "Lower River Colne evaluation report," (November 1989).
5. National Rivers Authority Thames Region. "River Stort evaluation report," (November 1989).
6. National Rivers Authority Thames Region. "Marsh dykes evaluation report," (March 1990).
7. Laurenson, E.M., and R.G. Mein. RORB — Version 3 Runoff Routing Program User Manual, 2nd ed., Monash University (March 1985).
8. Evans, E.P., and V. Lany. "A mathematical model of overbank spilling and urban flooding," *International Conference on the Hydraulic Aspects of Floods and Flood Control* (London 1983).
9. National Rivers Authority. "Corporate Plan 1990/91," (September 1990).
10. National Rivers Authority Thames Region. "Business Plan 1991/92," (November 1990).
11. National Rivers Authority Thames Region. "River Stort draft catchment management plan," (April 1991).
12. Tydac. "Spans spatial analysis system version 4.3," (1988).

POLLUTION TRANSPORT
AND CONTROL

5 ASSESSMENT OF STORMWATER POLLUTION TRANSPORT TO SWEDISH COASTAL WATERS

5.1 INTRODUCTION

Studies dealing with stormwater pollution transport on a local scale have been carried out in Sweden since the middle of the 1970s.[1-7] However, so far no investigation has aimed at determining the importance of stormwater pollution on a larger scale, involving a region or a part of the country. Such studies would provide the necessary background information for (1) developing a regional (nationwide) stormwater management policy, (2) quantifying the importance of stormwater pollution in comparison with other pollution sources, and (3) helping to set the priorities between different measures taken to reduce the pollution transport to receiving waters.

The pollution transport in Sweden through stormwater runoff is of great significance for many inland receiving waters, and thus must be considered when measures for water quality enhancements are discussed. In this context, questions have been raised as to what degree stormwater pollution contributes to the water quality conditions in the coastal waters surrounding Sweden. The Swedish coastal waters are normally divided into six different water bodies, namely the Bothnian Bay, the Bothnian Sea, the Baltic Sea, the Sound, Kattegat, and Skagerrak. Figure 1 displays the geographic extent of these coastal water bodies and the different countries which are bordering the water bodies.

The continuous deterioration of the water quality in these coastal waters is a matter of great concern, and extensive measures are planned to reduce the total pollution transport, especially with regard to nutrients, such as nitrogen and phosphorous. However, these measures focus on the pollution transport from nonpoint sources in areas of agriculture and forestry, and the contribution from stormwater runoff in urban areas has largely been neglected. It is not obvious that

Figure 1.   Location map of the study area and the division of Swedish coastal waters into six water bodies.

the pollution transport from urban areas through the stormwater is negligible regarding nutrients, and with respect to metals stormwater is typically one of the major sources for polluting receiving waters.

The objective of this study is to quantify the total pollution transport to Swedish coastal waters in connection with stormwater runoff. Calculations of the stormwater pollution transport were carried out for total nitrogen (N_{tot}), total phosphorous (P_{tot}), and several metals (Pb, Zn, Cu, Cd, and Cr). The calculated stormwater pollution transport was compared with the contribution from other pollution sources in Sweden such as agriculture, forestry, industry, and urban areas (excluding stormwater runoff). Comparisons are also made with reference to the pollution transport from other countries to the studied coastal receiving waters. An interesting management aspect of developing a strategy for pollution control, also involving stormwater runoff, is that it is a multinational problem that includes not only Sweden but Finland, Russia, Poland, Germany, Denmark, and Norway.

5.2 DEMOGRAPHIC AND LAND-USE CHARACTERISTICS IN SWEDEN

The total area of Sweden is about 450,000 km^2 of which 411,000 km^2 is land surface. The urban areas constitute approximately 5000 km^2 or 1.2% of the land

Figure 2.   Location and areal extent of the counties in Sweden (letter abbreviation denoting respective county given in the figure).

surface. More than 80% of the entire Swedish population, consisting of about 8.5 million people, is living in the urban areas, which are mainly located in the southern part of the country. The most densely populated urban areas are situated in the vicinity of the three largest cities in Sweden namely, Stockholm, Gothenburg and Malmöe (see Figure 1).

Sweden is divided into 24 counties or provinces for administrative reasons. Figure 2 illustrates the location and areal extent of these counties. In the figure are also the letter abbreviations given that are used to denote the respective counties. The present study was performed on a spatial scale corresponding to county level since many statistics in Sweden are reported for these geographic units. Furthermore, carrying out the study on a smaller scale would considerably increase the computational effort without improving the accuracy notably with the used methodology. Figure 3 displays the total amount of urban area and the population in each of the counties in Sweden.[8,9]

Land use in urban areas was classified into eight different categories that are presented in Table 1 together with the areal percentage respective land-use type encompasses as an average for all urban areas in Sweden. The term infrastructure includes facilities such as streets, railroads, and harbors. Undeveloped land refers to areas which are not possible to classify according to any other specific

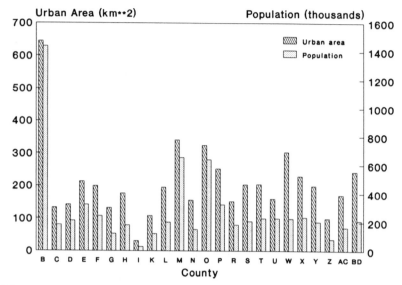

Figure 3. The total amount of urban areas and the population in respective county in Sweden.

Table 1. Areal Percentage of Different Types of Land Use in Urban Areas in Sweden, Measured as an Average for the Whole Country

Type of Land Use	Areal Percentage
Residential area	33
Industrial area	12
Commercial area	9
Infrastructure	14
Recreational area	2
Agricultural area	3
Forested area	14
Undeveloped	12

category. About 69% of the land use in urban areas involves activities where at least a part of the surface is paved, and thus will typically generate surface runoff in connection with rainfall. Land-use categories where little surface runoff is expected were recreational, agricultural, and forested areas together with undeveloped land, in total comprising approximately 31% of the urban areas.

The amount of runoff and the quality of the stormwater is dependent upon the land-use category. Residential areas with a high population density will, for example, have a considerably larger runoff coefficient than areas that are

Table 2. Representative Concentrations for N_{tot} and P_{tot}
for Different Types of Land Use[5]

Type of land use	N_{tot} (mg/L)	P_{tot} (mg/L)
Residential area (densely populated)	1.0–3.0	0.2–0.6
Residential area (sparsely populated)	1.0–2.5	0.2–0.4
Industrial area	1.5–2.5	0.2–0.6
Infrastructure	1.5–2.5	0.2–0.4

sparsely populated. The runoff coefficient, expressing the amount of net rainfall that will occur as stormwater runoff, typically varies between 0.10 and 0.50 for larger urban areas. For small urban areas in the center of cities considerably higher runoff coefficients are possible since most of the surface is paved. Similarly, the typical concentration in the stormwater for a specific pollutant is intimately linked to the type of land use. Representative concentrations in the stormwater have been presented for several different pollutants and land use type based on extensive studies[5] in various urban areas in Sweden. Table 2 summarizes some results from these investigations regarding representative concentrations for N_{tot} and P_{tot} for different types of land use.[5]

The instantaneous concentration in the stormwater during a runoff event is naturally highly variable and dependent on factors such as amount of pollutants accumulated on the runoff surfaces, rainfall intensity, and duration. However, representative concentrations often provide a good description of the pollution transport over longer periods of time, and could be effectively used in calculations performed to yield a foundation for taking measures to reduce stormwater pollution.

5.3 CALCULATION OF STORMWATER POLLUTION TRANSPORT

In order to quantify the pollution transport to Swedish coastal waters in connection with stormwater runoff, calculations were carried out for each county separately. Every county was assigned a typical annual amount of precipitation, based on historical data, varying between 550 and 900 mm. An annual rainfall loss of 150 mm, caused by depression storage, was subtracted from the annual precipitation to obtain the net rainfall. The urban areas within each county were divided into three groups depending on population, that is, class I: $P > 100,000$, class II: $100,000 > P > 20,000$, and class III: $20,000 > P$, where P denotes the population. Different runoff coefficients and characteristic concentrations were chosen for each class. These concentrations were selected based upon studies

Table 3. Runoff Coefficients and Characteristic Concentrations for
Urban Areas of Different Size in Sweden

Class	Runoff Coeff.	N_{tot}	P_{tot}	Pb	Cu (mg/L)	Zn	Cd	Cr
I	0.30	2.5	0.45	0.15	0.12	0.30	0.003	0.008
II	0.20	2.0	0.35	0.10	0.08	0.20	0.002	0.006
III	0.15	1.5	0.25	0.05	0.04	0.10	0.001	0.004

dealing with stormwater quality[5], involving urban areas of varying size. The representative values are summarized in Table 3.

The runoff volume was determined for each urban area within a county from the runoff coefficient, the net rainfall amount, and the size of the urban area. The pollution load for the urban area was calculated from the runoff volume and the representative concentration, and then summed to obtain the total pollution load from the entire county. Demographic data provided the necessary information to determine to which class a specific urban area belonged, and thus what representative values to apply.

The stormwater discharge and pollution transport to the different coastal waters were calculated by including the counties that contributed to a specific coastal water. When carrying out this summation, it was assumed that a certain reduction in the pollution transport occurred for stormwater discharged to an inland receiving water due to chemical and biological processes. All metals and P_{tot} were reduced with 25%, whereas N_{tot} was reduced with 50%. The pollution transport from urban areas discharging stormwater directly to coastal waters was not reduced. In the calculations it was assumed that the stormwater was discharged to the receiving water without any treatment. However, in some urban areas with combined sewers, stormwater is discharged together with the wastewater, and thus a portion of the stormwater may undergo treatment during a rainfall. No modifications of the pollution transport calculations were made with respect to possible treatment of the stormwater. In combined systems, the stormwater in general causes an increase in the pollution transport to the receiving waters due to combined sewer overflow.

5.4 RESULTS

5.4.1 Stormwater Pollution Transport

The average annual stormwater runoff in Sweden was estimated at 500 million m^3, of which approximately 240 million m^3 is discharged directly to the coastal waters. The rest of the stormwater runoff is transferred to the coastal waters through inland receiving waters, such as rivers and lakes experiencing the aforementioned reduction in pollution transport. Table 4 presents the average

Table 4. Average Annual Pollution Transport by Stormwater
in Sweden for Various Pollutants (Total Amount
Generated at the Source of Stormwater Runoff)

Pollutant	Total Amount (kg/year)
P_{tot}	160,000
N_{tot}	920,000
Pb	42,000
Cu	34,000
Zn	85,000
Cd	850
Cr	2,700

annual pollution transport by stormwater in Sweden for the studied pollutants. The pollution transport refers to the total amount generated in connection with stormwater runoff and involves no reduction due to biological and chemical reactions.

As expected, the calculations showed that the counties where the major urban centers are located dominate the stormwater pollution transport. Figure 4 illustrates the annual transport of N_{tot} and P_{tot} by stormwater from each of the counties in Sweden. The counties where the three major cities in Sweden are located (Stockholm, Gothenburg, and Malmoe) contribute more than 40% of the total pollution transport by stormwater for N_{tot} and P_{tot}. Furthermore, the county where Stockholm is situated has an N_{tot} and P_{tot} stormwater pollution transport corresponding to more than 20% of the total transport from Sweden. Similar ratios as those observed for the nutrients are applicable for the metals studied (see Table 4).

5.4.2 Pollution Transport To Swedish Coastal Waters

The annual pollution transport by stormwater to the Swedish coastal waters are 140,000 kg P_{tot}, 710,000 kg N_{tot}, 38,000 kg Pb, 30,000 kg Cu, 76,000 kg Zn, 760 kg Cd, and 2400 kg Cr, where the transport was reduced as previously discussed. Table 5 summarizes the annual stormwater pollution transport to the different coastal waters (Figure 1) for the pollutants studied. The Baltic Sea is the largest recipient for stormwater discharge and receives 40 to 45% of the total transport from Sweden depending on the pollutant. The smallest pollution transport with the stormwater is to the Sound, which geographically also has the smallest extent.

Since the coastal waters have different areal extent, the stormwater pollution transport should be compared to the contribution from other pollution sources in order to determine the importance of stormwater runoff as a pollution source for respective coastal water. Such a comparison could provide the basis for setting

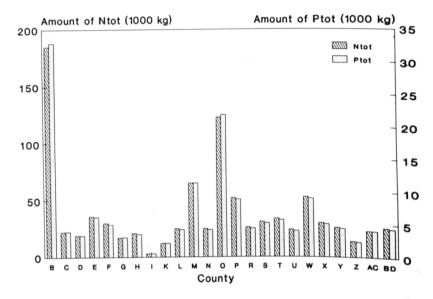

Figure 4.   Annual pollution transport of N_{tot} and P_{tot} by stormwater for respective counties in Sweden.

Table 5.   Annual Stormwater Pollution Transport to the Different Coastal Waters in Sweden

	Volume (Million m³)	P_{tot}	N_{tot}	Pb (kg)	Cu	Zn	Cd	Cr
Bothnian Bay	26	6,700	35,000	1,500	1,200	3,100	31	110
Bothnian Sea	88	22,000	105,000	5,300	4,300	10,700	110	360
Baltic Sea	205	60,000	305,000	17,000	13,000	34,000	340	1,000
The Sound	20	6,500	35,000	1,800	1,500	3,700	37	110
Kattegat	125	32,000	149,000	8,100	6,500	16,100	160	530
Skagerrak	37	14,000	78,000	4,400	3,500	8,800	88	250

the priorities between different measures taken to reduce the pollution load to the coastal waters. Figure 5 depicts the contribution from the stormwater to the different coastal waters presented for P_{tot} and N_{tot} expressed in percentage of the total pollution transport from Sweden. The pollution transport from other sources to Swedish coastal waters was obtained from the Swedish Environmental Protection Board,[10] and includes transport from agriculture, forestry, industry, and various municipal sources besides stormwater such as treated wastewater.

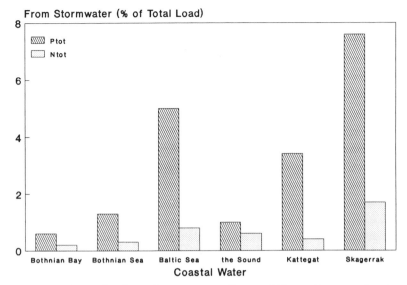

Figure 5.   Contributions of stormwater, P_{tot}, and N_{tot} to the coastal waters as a percentage of the total pollution transport from Sweden.

As seen from Figure 5, the stormwater is of greater importance for the total transport of P_{tot} than for the transport of N_{tot}. The transport of P_{tot} through the stormwater is of greatest significance for Skagerrak, Kattegat, and the Baltic Sea. In Skagerrak the stormwater contributes to almost 8% of the total transport of P_{tot} from Sweden, mainly due to the large percentage of urban areas at the West coast of Sweden and the large amount of rainfall in this area. The Baltic Sea, which receives the largest transport of P_{tot} and N_{tot} through the stormwater in absolute numbers, is the second largest recipient of P_{tot} and N_{tot} in terms of percentage of the total transport. The contribution from the stormwater to the total transport of N_{tot} from Sweden to the different coastal waters is below 2%, and in all cases except for Skagerrak below 1%.

The importance of the pollution transport by stormwater for metals is more difficult to asses, since reliable estimates of the transport of metals from Sweden to the coastal waters are lacking to a large extent. However, one study[11] presented an estimate of the transport of metals to the coastal waters surrounding Sweden (except for Skagerrak), including the pollution transport from all countries bordering these coastal waters. The evaluated metal transport encompassed coastal and river input together with deposition from the atmosphere and wastewater effluents. As seen in Figure 1, pollutants are transported to the studied coastal waters not only from Sweden, but also from Finland, Russia, Poland, Germany, Denmark, and Norway. Urban stormwater runoff is a major pollution source mainly with respect to the transport of Pb, where the stormwater metal transport from Sweden amounts to 1 to 6% of the total transport[11]

Table 6. The Contribution from Sweden in Terms of the Total Pollution Transport (Not Including the Stormwater) to the Different Studied Coastal Waters for P_{tot} and N_{tot}

	P_{tot} (%)	N_{tot} (%)
Bothnian Bay	36	33
Bothnian Sea	46	38
Baltic Sea	2	4
The Sound	6	6
Kattegat	25	42
Skagerrak	5	5

depending upon the coastal recipient. For the other metals, the transport through the stormwater in general constitutes less than 1% of the total transport from all the contributing countries.

5.5 CONCLUDING DISCUSSION

The main objective of this paper was to assess the importance of stormwater pollution transport from Sweden to the coastal waters, especially in comparison with other pollution sources such as agriculture, forestry, and industry. Such estimates are of value for setting priorities between different measures to reduce the pollution transport to Swedish coastal waters, and thus to improve water quality. Previous investigations on the pollution transport to Swedish coastal waters have not included stormwater, but the only studied pollution sources from urban areas are effluents from wastewater treatment plants.

An additional complication regarding water quality improvement in Swedish coastal waters is that pollutants are transported from several countries. Thus, even if the pollution transport from Sweden is significantly reduced, and if no measures are taken in other countries bordering Swedish coastal waters, limited water quality improvement is expected. The contribution from Sweden in terms of the total pollution transport (not including stormwater) to the different coastal waters has been presented[10] for P_{tot} and N_{tot} (see Table 6). Sweden is a major contributor regarding the nutrient transport to the Bothnian Bay, Bothnian Sea, and Kattegat, whereas the Baltic Sea receives limited amounts compared to the transport from other countries.

In summary, the stormwater pollution transport could constitute a significant portion of the total transport from Sweden for certain pollutants and coastal waters. For example, the marked transport of P_{tot} and Pb in the stormwater indicates that reduction or treatment of urban stormwater runoff could be a beneficial method to improve water quality in some Swedish coastal waters. However, a comparison between the pollution transport from Sweden and from other countries bordering Swedish coastal waters indicates that even extensive reduction of the transport from Sweden will have little effect on the water quality in most of the coastal waters. Thus, to improve water quality in Swedish coastal waters extensive measures are needed also in other countries bordering these water bodies.

REFERENCES

1. Malmquist, P.-A. "Heavy metals in urban storm water," in *Proceedings of the International Conference on Heavy Metals in the Environment* (Toronto, Canada, 1975).
2. Malmquist, P.-A., and G. Svensson. "Urban storm water pollution sources," in *Proceedings of the Amsterdam Symposium*, IAHS Publication No. 123 (1977), p. 31.
3. Hogland, W., and J. Niemczynowicz. "Quantitative and qualitative water budget for the city of Lund," Report No. 3029, Department of Water Resources Engineering, Lund Institute of Technology, University of Lund, Lund, Sweden (1979), in Swedish.
4. Hogland, W., and R. Berndtsson. "Quantitative and qualitative characteristics of urban discharge to small river basins in the South West of Sweden," *Nordic Hydrol.* 156 (1983).
5. Malmquist, P.-A. "Urban stormwater pollutant sources. An analysis of inflows and outflows of nitrogen, phosphorus, lead, zinc and copper in urban areas," Chalmers University of Technology, Göteborg, Sweden (1983).
6. Hogland, W. "Rural and urban water budgets. A description and characterization of different parts of the water budgets with special emphasis on combined sewer overflows," Report No 1006, Department of Water Resources Engineering, Lund Institute of Technology, University of Lund, Lund, Sweden (1986).
7. Svensson, G. "Modelling of solids and metal transport from small urban watersheds," Chalmers University of Technology, Göteborg, Sweden (1987).
8. SCB. "Urban areas in Sweden," Report NA21-SM-8601, Statistics Sweden, Örebro, Sweden (1986).
9. SCB. "Land Use in Urban Areas in Sweden," Report NA14-SM-8701, Statistics Sweden, Örebro, Sweden (1987).
10. Enell, M. "Impact of utilization in river basin management," *Aqua Fennica* 19: 153 (1989).
11. Lithner, G., H. Borg, U. Grimås, A. Göthberg, G. Neumann and H. Wrådhe. "Estimating the load of metals to the Baltic Sea," *Ambio*, 7: 7 (1990).

6 A LEVEL SPREADER/ VEGETATIVE BUFFER STRIP SYSTEM FOR URBAN STORMWATER MANAGEMENT

6.1 INTRODUCTION

A vegetative buffer strip (VBS) is a stormwater management practice which utilizes a vegetated surface to reduce runoff velocities, allow infiltration, and filter out runoff pollutants. The VBS has been found to be a cost-effective control measure of reducing solids, nutrients, and other pollutants in stormwater runoff. For example, Barfield and Albrecht reported that a vegetative buffer strip could remove approximately 70 to 99% of all fine grained sediment.[1] More recently, Dillaha et al.[2] monitored VBS's on 33 farms in Virginia, to evaluate their long-term effectiveness and water quality improvement. Among their findings, the most significant factor affecting the performance of vegetative buffer strips was the flow regime. It concluded that when water was allowed to achieve a concentrated flow profile, only minor pollutant reduction was achieved. However, good removal efficiencies for suspended solids and particulate nutrients were obtained when overland flows were spread evenly across the filter strips.

Laboratory research was conducted at the University of Kentucky on simulated grasses in order to gain a stronger understanding of the filtration dynamics of grasses.[3] Conclusions that grass filters provide excellent trapping efficiencies and that the establishment of grass filters is much more economical than the use of detention basins. A later laboratory study by Kao concluded that continuous flooding did not appear to reduce filtration efficiencies and that longer filter lengths would be needed if colloidal size particles were to be removed.[4]

This paper describes a novel approach to improving the pollutant removal efficiency and the reliability of a VBS by using a level spreader (LS). The design

0-87371-805-4/93/$0.00+$.50

93

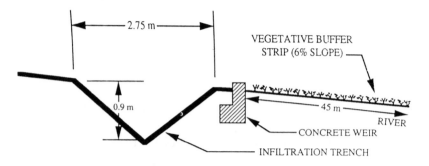

Figure 1.   Details of a level spreader.

of such an LS/VBS system is described and results from a full-scale field test of such a system are summarized.

6.2 THE LEVEL SPREADER/VEGETATIVE BUFFER STRIP SYSTEM

6.2.1 General Description

The level spreader is essentially an earthen trench placed on the contour and has a concrete weir spillway on the downslope side of the trench (Figure 1). Thus, the level spreader takes the concept of a vegetative buffer strip one step further by combining it with an infiltration trench and a concrete weir spillway for an enhanced method of stormwater pollution control. Referring to Figure 1, when a storm occurs, the discharge is routed into an infiltration trench, which serves as a miniature detention basin. Water fills this trench, and then flows over the concrete weir built in front of the trench and continues downslope through the vegetative buffer strip. Removal of stormwater pollution takes place in both the infiltration trench and in the vegetative buffer strip.

Mechanisms associated with the level spreader/vegetative buffer strip system can be summarized as the following:

- settling which occurs in the level spreader itself
- sedimentation and filtration, removing primarily solids and metals by the filter strip
- adsorption, precipitation, and plant uptake, removing primarily nutrients by the filter strip

6.2.2 The Field Experimentation Site

A level spreader/vegetative buffer strip system was built near a shopping center east of Charlottesville, Virginia. The watershed that drains into the system

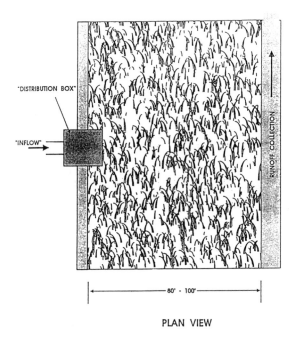

PLAN VIEW

Figure 2.   Plan view of the level spreader/buffer strip system.

is a 4-ha commercial area with almost 100% imperviousness. Referring to the system plan view depicted in Figure 2, runoff from the shopping mall is collected by a 30-in. (76.2-cm) storm sewer which drains into a trapezoidal channel. It continues downslope until it reaches a centrally located distribution box which serves to divert the storm runoff to the level spreader trenches. The triangular-shaped infiltration trenches are constructed perpendicular to the slope on a zero percent grade and are approximately 9 ft wide by 3 ft deep (2.5 m x 0.9 m). Runoff from the shopping mall fills these trenches and then flows over a concrete weir built in front of the trenches (Figure 1). From here it continues downslope through the vegetative buffer strip. The drainage system which includes the level spreader has been designed to handle the peak runoff from a 10-year return period, 10-min storm. The peak discharge from the mall outlet pipe is estimated at approximately 60 cfs or 1.70 cms. The distribution box was constructed to divert 60% of the peak flow, or 1.02 cms into the level spreader and 40% of the peak flow into an emergency spillway, which drains into the Rivanna River.

The level spreaders have a total length of approximately 170 m, a cross-sectional area of 2.30 m^2, and a total volume capacity of 2,140 m^3. The makeup of soils at the site was 15% clay, 19% silt, 20% fine sand, 24% coarse sand, and 22% gravel. The average permeability was about 1.0E-07 cm/sec. The grass type was a Kentucky 31 that covered an effective area of about 3770 m^2.

Table 1. Storm Events Sampled in 1987

Date	Time Sampled	Total Approx. Precip. for Day (in./mm)
March 30	10:50–17:05	0.68 (17.3)
April 15	11:00–14:40	1.28 (32.5)
April 16	16:00–17:00	3.74 (95.0)
April 17	10:00–11:00	0.61 (15.5)
May 13	17:05–17:45	0.02 (0.5)
May 19	20:45–23:15	0.46 (11.7)
June 9	16:00–17:00	0.53 (13.5)
June 13	16:35–16:55	0.43 (10.9)

6.2.3 Sampling Scheme

With the site properly prepared, a sampling scheme was devised which would lend itself to uniformity in collection and also prove to be most effective. It was decided to sample from three general locations. They were at the outlet from the shopping mall, at the level spreader itself, and downslope from the level spreader. Samples at the mall outlet were grabbed by hand while the samples from the level spreader were taken by an automatic sampling device and also by hand. For overland flow sampling, a collection trough and turkey baster method was used. Troughs approximately 12 in. (30 cm) long and triangular in shape (volume of 500 cc) were used. These troughs were of lightweight design and of sturdy aluminum construction and were placed at various distances downslope of the level spreader for sampling the entire extent of overland flow. During the period from March through June, 1987, eight storm events were sampled. The dates, time, and rainfall data are presented in Table 1.

At each sampling location, triplet water samples were taken and analyzed to determine the concentrations of total suspended solids, nitrate/nitrite, total phosphorus, lead, and zinc. The analytical procedures used to measure pollutant concentrations were from the U.S. EPA *Methods for Chemical Analysis of Water and Wastes*.[5]

To further describe the field setup of the LS/VBS system, two photos are presented here. Figure 3 shows storm runoff flows out from the 30-in. sewer down to the distribution box. Figure 4 depicts the level spreader and the grass strip. It can be seen that the level spreader is about full. The wooden box is for storing water samplers and other equipment.

6.3 RESULTS AND DISCUSSION

The removal efficiency of the LS/VBS system was estimated by the mass balance approach, i.e.,

Figure 3.   Open channel conveying storm runoff to distribution box.

Figure 4.   Level spreader with vegetative buffer strip.

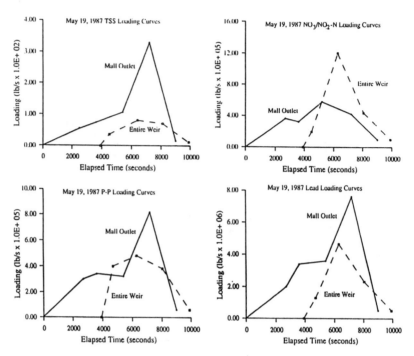

Figure 5.   Examples of pollutant mass loading curves.

$$(\text{Mass into the system}) - (\text{Mass out from the system}) = (\text{Mass removed})$$

Pollutant mass or loading was calculated by multiplying the observed concentration and the corresponding flow rate.

Before a monitoring system could be devised for accurate measurement of level spreader and overland flow rate, a first attempt was made to obtain pollutant loadings by using the available field data.

Essentially, it was assumed that the level spreader was perfectly "level" and therefore any overflow would occur uniformly along the entire length of the level spreader. The flow rate over the experiment strip could then be apportioned by a factor which was equal to the ratio of the experiment strip length to the level spreader length. It was further assumed that a uniform overland flow occurred downslope from the level spreader weir. Of course, flow rate would actually decrease downslope due to storage, infiltration, and retardation by the grass.

Loading rates for various pollutants at specific sampling locations and times were then computed from known concentrations and flow rates. Examples of the computed pollutant loading curves are shown in Figure 5.

Based on the mass balance principle, removal efficiencies of the LS/VFS system for four complete storm events were calculated and listed in Table 2. The results indicate the following:

**Table 2. Pollutant Removal Efficiency (%) LS/VFS System
Distance Downslope from Level Spreader (m)**

Storm	Pollutant	6	12	16	21
45 (end)					
	TSS	41		43	
	NO_3+NO_2	28		29	
3/30/87	T–P	47		45	
	PB	28		33	
	ZN	24		10	
	TSS	21	49		54
70					
	NO_3+NO_2	(34)[a]	(3)		(27)
4/15/87	T–P	15	10		25
35					
	PB	(55)	18		(16)
	ZN	26	40		55
49					
	TSS	40	48		84
	NO_3+NO_2	0	8		40
5/19/87	T–P	24	18		25
	PB	9	22		39
	ZN	22	28		50
	TSS	9	23		73
72					
	NO_3+NO_2	12	14	20	20
	T–P	6	13	33	40
6/9/87	PB	27	15	38	50
	ZN	0	5	47	53

[a] Numbers in parenthesis indicate negative removal rates.

- The LS/VBS system is effective in removing particulate pollutants such as suspended solids, and is not as effective in removing soluble pollutants such as nitrate and nitrite.
- Removal efficiency increases as VFS length increases.

Results obtained regarding detention and VFS pollutant removal efficiency can be summarized in the following sections.

Table 3. Comparison of Bottom and Surface Pollutant Concentrations in the Level Spreader Concentrations in mg/L

Storm		Solids	NO$_3$+NO$_2$	Total-P	Lead	Zinc
4/16/87	Bottom	1054	0.01	1.17	0.10	0.32
	Surface	57	0.02	0.18	0.02	0.05
4/17/87	Bottom	56	0.23	0.13	0.02	0.07
	Surface	18	0.20	0.10	0.02	0.07
5/19/87	Bottom	83	0.40	0.40	0.03	0.20
	Surface	50	0.52	0.34	0.02	0.18

6.3.1 Detention at the Level Spreader

For several storm events, samples were taken both at the bottom and at the surface of the level spreader. Results are given in Table 3.

The results show a high settling rate for suspended solids, and significant settling for phosphorus, lead, and zinc for some storms. On the other hand, settling is poor for nitrate and nitrite.

6.3.2 Vegetative Filter Strip Removal

As shown in Table 2, the following removal efficiencies were obtained for filter length 21 m and 45 m based on the limited data collected:

Pollutant	Removal Efficiency, Percent
	21m–45m
Suspended Solids	54–84
NO$_3$+NO$_2$	(27)–20
Total phosphorus	25–40
Lead	(16)–50
Zinc	47–55

These results show that removal rates are good for solids, phosphorus, and zinc, poor for nitrate/nitrite and lead. The negative rates were probably due to soil erosion of the hillslope. When grass is denser and loss of runoff due to depression and detention is accounted for, the removal rates would be higher.

6.3.3 Effect of Filter Length and Design Implications

Overall pollutant removal rates for the four complete storm events were computed by averaging the rates and are listed according to filter length in Figure 6.

The results suggest that filter length is an important design parameter, as mentioned previously in the literature.[2]

It appears, by examining Figure 6, that a filter with a length of 70 ft or 25 m would provide good removal when compared to longer strips.

6.3.4 Consideration of Cost Effectiveness

The pollutant removal efficiencies of a wet pond and the LS/VFS system were compared,[6] as listed below:

Overall Pollutant Removal Efficiency, Percent

Pollutant	Wet Pond	Pantops LS/VFS S (Full VFS Length)
Suspended solids	77	71
Total phosphorus	70	38
Nitrate-nitrite	75	10
Lead	57	25
Zinc	50	51

The above results suggest that the LS/VFS system is as effective as the wet pond in removing solids and zinc, but not so for phosphorus, lead, and nitrate-nitrite. It should be noted however, that a dense turf was not in place when the storm events were sampled and erosion did occur downslope from the LS weir. It is expected that a properly constructed and maintained LS/VFS system will be most effective during warm weather months when vegetation is in full, vigorous growth. It should be noted also that receiving waters are most vulnerable to pollutant inputs during warm-weather months.

The level spreader was built at a 1986 cost of $15,420 and services approximately 4 ha of essentially impervious drainage area. On the other hand, a typical detention basin, such as a regional detention pond in Charlottesville, Virginia, costs approximately $60,000, while servicing approximately 16 ha of mixed land use. The payoff is that in exchange for the level spreaders performance and cost, maintenance must be performed more often.

It appears that the LS/VFS system is at least as cost effective as a wet pond in removing pollutants in storm runoff. It should be noted that the LS/VFS system does not require a large amount of land and is particularly suitable for a hillslope topography.

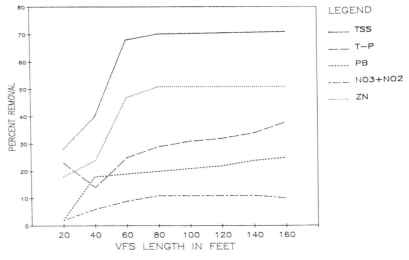

Figure 6.   Removal efficiency vs. vegetative strip length.

6.3.5 Mathematical Model for the Level Spreader System

The ability to predict the performance of level spreaders as a best management practice is an essential component in the design and analysis of such a system. The use of mathematical flow and pollutant transport/transformation models is well established in both surface and subsurface water quality management. Mathematical models offer the advantage of analyzing various design options and thus allowing the selection of an optimal design compatible with site conditions and pollutant loading constraints. Although numerous flow and transport models have been developed for surface and subsurface systems separately, little attention has been given to the coupled surface, subsurface flow, and transport problem characteristics of a level spreader.

A coupled surface and subsurface flow and pollutant transport model is being developed. The model has been formulated and a four-point implicit numerical scheme is being used for solving the transport equations.

A detailed documentation of the model development process and a review of the existing literature on the subject are given in Byearne and Yu.[8]

REFERENCES

1. Albrecht, S.C., and B.J. Barfield. "Use of a vegetative filter zone to control fine grained sediments from surface mines," paper presented at the 1982 Symposium on Surface Mining Hydrology, Sedimentology, and Reclamation, University of Kentucky, Lexington, KY, 1982.

2. Dillaha, T.A., J.H. Sherrard and D. Lee. "Long-term effectiveness and maintenance of vegetative filter strips," Bulletin 154, Virginia Water Resources Research Center, Virginia Polytechnic Institute and State University, Blacksburg, VA, 1986.

3. Kao, D.T., B.J. Barfield and A.E. Lyons. "On-site sediment filtration using grass strips," paper presented at the National Symposium on Urban Hydrology and Sediment Control, University of Kentucky, Lexington, KY, July 1975.

4. Kao, D.T. "Determination of sediment filtration efficiency of grass media, Volume 1: Sediment filtration efficiency of continuous grass media," Research Report No. 124, Project No. A-069-KY, University of Kentucky, Water Resources Research Institute, Lexington, KY, 1980.

5. "Methods for chemical analysis of water and waste," U.S. EPA Report-600/4-79-020, Environmental Research Center, Cincinnati, OH (1979).

6. Yu, S.L., W.K. Norris and D.C. Wyant. "Urban BMP demonstration project in the Albemarle/Charlottesville area," Report No. UVA/530358/CE88/102, Grant no. A85-3798(06581) (1987).

7. County of Albemarie, Engineering Office. Personal communication (1987).

8. Byearne, M.R., and S.L. Yu. "Coupled overland-subsurface flow and transport modeling," Proceedings of the International Conference on Computer Applications in Water Resources, Taipei, Taiwan (1991), p. 2844.

7 TRANSITION OF POLLUTANT RUNOFF FROM A SMALL RIVER BASIN IN URBAN AREA UPON SEWERING

7.1 INTRODUCTION

Because of the multiple uses of water, the high potential for pollution, complex land uses, and sewer systems, comprehensive environmental water management, which includes all of these factors, is becoming necessary in urban areas. Therefore, from the characteristics of pollutant runoff during a storm event, which is one of the most significant factors in comprehensive environmental water management in urban areas but has not yet been explained sufficiently, this paper is aimed to clear the following three subjects related to receiving water:

1. quantification of the effect of sewering in an urban river basin
2. the pattern of pollutant runoff including the first-flash
3. evaluation of the pollutant runoff from nonpoint sources

The pollutant load reduction due to sewering depends on both the decrease in discharge and the improvement in water quality. The long-term examination of these transitions in both discharge and water quality with the progress of sewering is important to quantify the effect of sewering. Up to this time, however, these examinations were limited mainly to the dry-weather period and have seldom been carried out during storm events taking a large portion of the annual runoff load.

In this paper, the pollutant load in a small river running through an urban area is examined, considering the urbanization of the river basin in terms of the progress of sewering and sewer types. The study area is the basin of the Tenjin River running through the northwest part of Kyoto City. It is a branch of the Yodo River used as a water resource for more than 10 million people in the Kinki area of Western Japan.

0-87371-805-4/93/$0.00+$.50
© 1993 by Lewis Publishers, Inc.

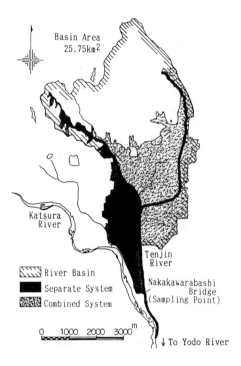

Figure 1.   Tenjin River Basin and sewer system.

Figure 1 shows the Tenjin River basin and the sewer system. The Tenjin River has an area of 25.7 km^2, and about 150,000 people live there. The basin has a relatively clear pattern of land use; the upper basin is mountainous or residential, the middle is residential, and the lower is industrial or residential. In particular, the industrial district in the lower basin is famous for its textile industry containing many of the factories in Kyoto City. The sewer system employed in the basin is separate in the right bank, and is combined in the left bank.

Table 1 shows the transition of social indices in the basin. Population, the residential area and the impervious area have changed very little during the last decade. However, the sewer system was constructed quickly during that period. Considering that it takes two years to connect with each house after the trunk line construction, the sewered rate in the population base in the combined area already achieved 100% in 1984, and in the separate area it reached 66.0% in 1989. Therefore, the ratio of pollutant load from non-point sources is becoming relatively higher.

Both data sets, which were obtained by continuous water sampling of the Tenjin River during the dry-weather period or during a storm event in the last decade and reported monthly by a local government, were analyzed considering the sewered rate. Tables 2a and b show the summary of the investigations.

Table 1. Transition of Social Indices

Year	Population ($\times 10^3$ Capita)	Residential Area for A_u(%)	Impervious Area for A_u(%)	Sewered Rate in Population Base (%)		
				Combined	Separate	Overall
1980	157.8	21.3	31.5	26.5	0.0	15.3
1984	154.6	21.8	32.3	100.0	22.5	67.3
1989	153.3	22.5	32.6	100.0	66.0	85.5

Note: Basin area; A = 25.75(km²), Urban area; A_u = 13.79(km²).

7.2 TRANSITION OF POLLUTANT RUNOFF DURING DRY-WEATHER PERIODS

The runoff load has been cut by the sewer system. Figure 2 shows the transition of discharge and pollutant load in the dry-weather period due to the sewered rate. The figure shows the discharge and pollutant load to be 75% of the unexcess value. In the last decade, the discharge decreased by 69%; the water quality decreased by 71% for T-COD, 43% for T-N, 78% for NH_4-N, and about 83% for T-P; while the sewered rate increased by 71%, from 15% to 86%. Consequently, the runoff load decreased by 87% for T-COD, 82% for T-N, 94% for NHT_4-N, and about 91% for T-P, i.e., noticeably in the pollutants affected by human actions. As SS is irregular and varies greatly, the reduction is not clear.

Dividing the last decade into two terms, Term I (1980 to 1984) and Term II (1985 to 1989), the result of the discharge and pollutant load investigations during the dry-weather period, which were averaged for weekdays and holidays respectively, were examined (Figure 3). Because little difference could be seen between weekdays and holidays in Term II, both were combined and averaged. There was a difference of about 1.2 to 1.5 times during the daytime in Term I, which was not seen in Term II. This resulted from the reduction in pollutant load from point sources in the daytime.

7.3 TRANSITION OF POLLUTANT RUNOFF AT A STORM EVENT

Pollutant runoff during a storm event is controlled by conditions such as the characteristics of rain, and the continuous dry days or integrated precipitation before the event. Few studies have been done that estimate several storm events under different conditions using the same scale. The ratio of runoff does not change much annually (Figure 4) because of the interaction of the suppressive effect of combined systems and the progressive effect of separate systems.

Table 2a. Summary of Investigations for Dry Weather Period

Period No.	Sampling Date as the Beginning	Weekday or Holiday	Antecedent Dry Days[a] (d)	Antecedent Precipitation[b] (mm)	Sampling Period (h)	Discharge (m³/s)	Mean Concentration (mg/L)				
							TR	SS	T-COD	S-COD	
Ts-1	80/06/12	Weekday	2	50.0	24.0	1.27	259.0	19.1	23.57	19.10	
Ts-2	80/06/19	Weekday	9	26.0	24.0	1.10	289.0	19.4	29.94	23.20	
Ts-3	80/08/03	Holiday	3	142.0	12.0	0.54	155.3	3.4	8.63	7.89	
Ts-4	82/06/06	Holiday	2	99.5	24.0	0.91	—	4.4	8.92	8.33	
Ts-5	82/07/30	Weekday	2	106.5	24.0	1.72	—	13.4	21.09	14.72	
Ts-6	83/08/26	Weekday	3	89.0	24.0	0.15	—	52.5	61.93	—	
Ts-7	85/07/23	Weekday	1	80.0	24.0	1.78	200.7	9.1	7.0	5.71	
Ts-8	85/08/04	Holiday	13	0.0	24.0	0.66	168.1	6.0	6.86	5.44	
Ts-9	86/11/25	Weekday	33	1.5	24.0	0.12	289.6	6.9	14.83	12.99	
Ts-10	90/10/24	Weekday	10	9.0	24.0	0.65	165.8	14.2	6.39	4.92	

[a] Period less than 10mm/d of precipitation before the day.
[b] Integrated precipitation for the past 10 days.

Table 2b. Summary of Investigations for Storm Event

Event No.	Date	Weekday or Holiday	Antecedent Dry Days[a] (d)	Antecedent Precipitation[b] (mm)	Precipitation (mm)	Duration Time (h)	Sampling Period (h)	Discharge (m³/s)	Mean Concentration (mL)			
									TR	SS	T-COD	S-COD
T-1	80/07/08	Weekday	0	97.5	12.0	4.0	6.0	4.36	528.7	405.9	51.42	7.47
T-2	80/07/23	Weekday	3	97.0	12.5	4.0	4.0	6.57	522.4	328.8	59.41	15.21
T-3	80/07/30	Weekday	5	111.0	30.0	9.0	8.7	12.64	427.6	321.5	34.22	9.31
T-4	80/08/10	Holiday	10	3.0	3.0	2.0	5.9	1.67	368.7	221.0	51.13	11.49
T-5	82/05/14	Weekday	7	17.5	11.0	3.7	3.3	2.83	—	357.5	65.32	13.79
T-6	82/10/19	Weekday	25	0.0	7.5	5.0	6.2	2.13	—	258.5	44.88	11.06
T-7	85/09/03	Weekday	22	1.5	8.0	0.7	3.3	5.01	1084.6	879.8	145.84	15.21
T-8	86/06/06	Weekday	8	36.0	12.0	9.0	12.8	1.42	219.5	12.9	14.96	12.49
T-9	86/06/17	Weekday	0	22.5	23.0	2.8	3.0	10.95	590.8	480.5	43.10	6.80
T-10	86/09/10	Weekday	22	0.5	4.5	0.5	2.7	3.21	264.2	47.1	25.80	15.15
T-11	87/06/03	Weekday	10	1.5	18.5	1.3	7.3	3.60	712.2	511.9	55.91	47.59
T-12	87/10/17	Weekday	2	20.5	1.0	0.3	4.5	0.96	172.9	45.7	13.65	6.50
T-13	89/03/31	Weekday	5	91.5	4.5	1.8	3.7	1.30	191.5	41.8	—	—

[a] Period less than 10mm/d of precipitation before the day.
[b] Integrated precipitation for the past 10 days.

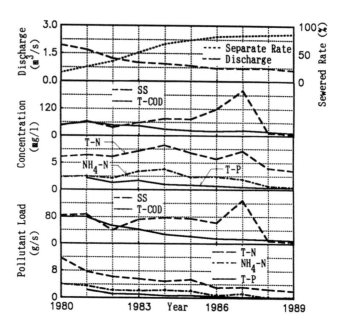

Figure 2.   Transition of discharge and pollutant load for dry-weather period.

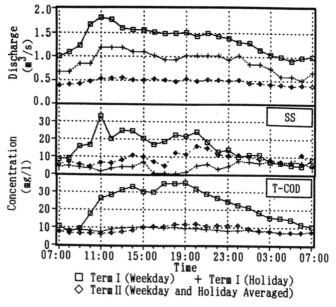

Figure 3.   Variation of runoff during a dry-weather day.

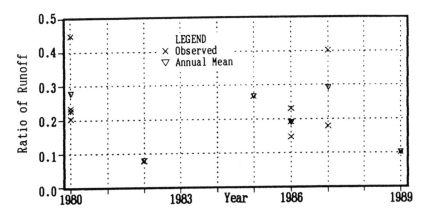

Figure 4.   Transition of ratio of the runoff.

Dr. Ebise showed that the integrated load during a storm event minus the load during the dry-weather period converged to a limiting value and that the integrated load and the integrated discharge could be formulated into Equation 1.

$$\Sigma L_{net} / A = a\left(\Sigma Q_{net} / A\right)^n \tag{1}$$

where $\Sigma Q_{net}/A$ and $\Sigma L_{net}/A$ are shown using the basin area (A), the integrated discharge or load observed during a storm event (ΣQ_{gross}, ΣL_{gross}) and the one during the same period in dry weather (ΣQ_{base}, ΣL_{base}) in Equations 2 and 3.

$$\Sigma Q_{net} / A = \Sigma Q_{gross} / A - \Sigma Q_{base} / A \tag{2}$$

$$\Sigma L_{net} / A = \Sigma L_{gross} / A - \Sigma L_{base} / A \tag{3}$$

Using Equation 1, each parameter a and n in Term I and II was determined by regression analysis. Table 3 shows the result, and Figure 5 shows the relationship between the integrated load and discharge. Although little difference could be seen in dissolved matter(DM), the regression line in each solid pollutant index, such as SS and T-COD, changed conspicuously from Term I to Term II. The two regression lines cross at the point due to 20 mm of precipitation considering the ratio of runoff. This means there is a significant effect of sewering on load reduction during a small storm event in Term II, but it does not mean that the effect is good during a large event. This is caused by the increase in the overflow load through a combined system and in pollutant runoff from nonpoint sources through a separate system.

Except for S-COD in Term I, the larger parameter n in the pollutant, which is considered to include more solid elements, means that the solid pollutants

Table 3. Regression Analysis Between Integrated Load and Discharge

Index	Term I			Term II		
	a	n	R^a	a	n	R^2
TR	1.5552	0.87	0.996	0.0030	1.66	0.912
SS	1.8125	0.83	0.977	0.0001	2.00	0.678
DM	0.5855	0.80	0.976	0.0657	1.12	0.917
T-COD	0.9420	0.66	0.996	0.0006	1.57	0.780
P-COD	1.0400	0.62	0.995	0.0006	1.45	0.480
S-COD	0.0093	0.99	0.756	0.0011	1.34	0.727

[a] R^2 = Coefficient of determination.

depend on discharge. Because the parameter n in each pollutant became larger from Term I to Term II, pollutants respond more sensitively to discharge.

As a result, it is inferred that pollutant loads would come to be more likely due to run off during annual storms. The effect of sewering is conspicuous only on solid pollutants during a small storm, but not so conspicuous on soluble pollutants or on every pollutant during a large storm event.

On the other hand, pollutant runoff at a storm event also depends on the potential pollutant load in the basin. Therefore, the change in the effect from the potential pollutant load is examined next. The integrated load in a storm was formulated into Equation 4, considering the continuous dry days or the integrated precipitation before an event as the potential factor.

$$\Sigma L_{net} / A = k \cdot S^m \left(\Sigma Q_{net} / A \right)^n \tag{4}$$

where the parameter S shows the potential of the pollutant load in the basin, using the continuous dry days before the event (with 10 mm of precipitation) as RUN-1 or the integrated precipitation before the event (for the antecedent 10 days) as RUN-2.

The result of regression analysis, except where $S = 0$, is shown in Table 4. Though relatively lower on SS and P-COD in Term II, as a whole, the coefficients of determination R^2 are improved so as to be higher than those in Equation 1. Generally, it is considered that the increase in continuous dry days before an event or the decrease in the integrated precipitation before an event causes runoff pollutant to increase, in which case $m > 0$ at RUN-1 and $m < 0$ at RUN-2 are presumed. However in Term I, the result was (1) $m < 0$ at RUN-1 except SS and (2) $m > 0$ at RUN-2 except S-COD, but in Term II, (1) m increased at RUN-1, except for SS, and (2) m decreased at RUN-2. This means that the effect of the potential pollutant load in the basin becomes conspicuous in pollutant runoff

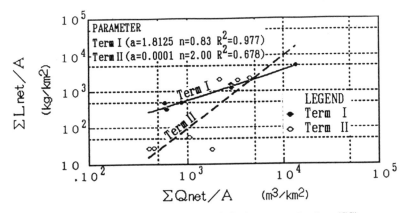

Figure 5a.   Result of regression analysis in integrated value (SS).

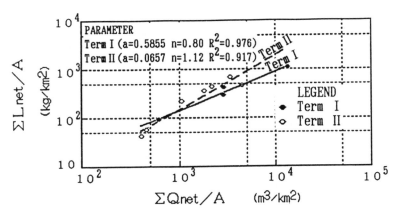

Figure 5b.   Result of regression analysis in integrated value (DM).

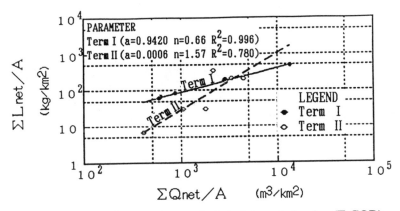

Figure 5c.   Result of regression analysis in integrated value (T-COD).

**Table 4 (1). Regression Analysis Between Integrated
Load and Discharge Considered Period less than 10 mm/d
in Precipitation before the Day < RUN-1 >**

Index	Term I				Term II			
	k	m	n	R[a]	k	m	n	R^2
TR	—	—	—	—	0.0028	0.19	1.62	0.905
SS	0.6299	0.22	0.91	0.993	0.0002	0.17	1.81	0.585
DM	—	—	—	—	0.0320	0.19	1.17	0.996
T-COD	1.4968	–0.09	0.62	1.000	0.0006	0.38	1.44	0.808
P-COD	1.1211	–0.01	0.61	0.997	0.0447	0.90	0.55	0.487
S-COD	1.6773	–1.07	0.59	0.933	0.00004	–0.08	1.87	0.941

[a] R^2 = Coefficient of determination.

**Table 4 (2). Regression Analysis Between Integrated
Load and Discharge Considered Integrated
Precipitation for the Past 10 days < RUN-2 >**

Index	Term I				Term II			
	k	m	n	R[a]	k	m	n	R^2
TR	2.1959	0.06	0.80	0.999	0.0074	–0.12	1.57	0.932
SS	1.5085	0.06	0.82	0.994	0.0004	–0.20	1.85	0.705
DM	0.8336	0.06	0.72	0.980	0.1447	–0.11	1.04	0.953
T-COD	1.2398	0.03	0.61	0.998	0.0011	–0.24	1.54	0.860
P-COD	1.2472	0.05	0.58	0.997	0.0011	–0.24	1.42	0.537
S-COD	0.0613	–0.08	0.81	0.930	0.0021	–0.25	1.31	0.839

[a] R^2 = Coefficient of determination.

during a storm event. Consequently, for most pollutant indices in Term II, the
results $m > 0$ at RUN-1 and $m < 0$ at RUN-2 have been obtained. For S-COD m
was negative at RUN-1, but this is because 10 mm of precipitation was selected
as the antecedent precipitation. It is inferred that much of this pollutant would run
off in a small storm even less than 10 mm of precipitation.

7.4 TRANSITION OF POLLUTANT RUNOFF THROUGH
  A YEAR

To estimate the yearly pollutant runoff related to the environmental management policy of water, the difference in the characteristics of pollutant runoff

between Term I and Term II was estimated by simulation using Equation 1 and Equation 4. Here, for Equation 4, the model was used that considered the precipitation integrated for ten days before the event as parameter S. SS, T-COD, P-COD, and S-COD were selected as the indices of pollutant. The time series of precipitation data per day observed by the Kyoto Regional Meteorological Observatory was used as input data for simulation after being transformed into effective rainfall (ΣQ_{net}/A). Three typical years during the last decade, 1980 (2041.0 mm of precipitation) as a flood year, 1984 (1170.5 mm of precipitation) as a drought year, and 1986 (1473.0 mm of precipitation) as a normal year, were selected for yearly simulation and applied to each equation in Term I and II.

Table 5 shows the result of simulation. The annual runoff load calculated by Equation 1 (Flow Type) in Term II sewered more than 80% is smaller than that by Equation 1 in Term I for every pollutant and for every year. The difference in S-COD, the dissolved pollutant, is small and the effect of sewering is not found. On the other hand, in the simulation by Equation 4 (Stock Type) through three years, flood, normal and drought, in spite of the small difference in S-COD between Term I and II as in Equation 1, the runoff load by Equation 4 in Term II is about 100 times as large as by Equation 4 in Term I for SS and P-COD, the solid pollutant, in the flood year 1980. It is presumed, in general, considering the accumulative load of pollutant in the basin, that the runoff load is not very different due to precipitation every year, relatively long period, assuming that the pollutant load originated in a year is not very different. However, the annual runoff load was calculated using Equation 4 as the large amount in proportion to the annual precipitation. Therefore it is considered that precipitation has a larger effect on runoff load than the accumulative load of pollutant. This is a characteristics of the Tenjin River.

The ratio of runoff load in a storm calculated by both equations in Term I to runoff load in a year, most of which (except S-COD) is close to 100%, is drastically decreased. Especially the ratio in Term II in the drought year decreased to nearly 20%. This means that in Term II sewered to a tolerable extent, the ratio of storm per year is reduced, but there is a large difference in runoff load due to the amount of precipitation. It is considered that the sensitivity of pollutant runoff to storms becomes greater because of sewering, though this is presumed from the much larger parameter n in all models in Term II than in Term I, and the pollutant from nonpoint sources, which is regarded to run off mainly during stormy weather days, occupies a large part of the runoff load in the sewered area. As a result, in an urban area sewered to some extent, one of the most important things is to control the pollutant runoff from nonpoint sources during stormy weather.

7.5 TRANSITION OF CHARACTERISTICS OF RUNOFF PEAK

The runoff load during a storm event is extremely concentrated. Generally, it is considered that the runoff load, in particular, is concentrated in: (1) the First-

Table 5. Simulation of Pollutant Runoff Through a Year

Water Quality	Year	Annual Precipitation (mm)	Term I (sewered rate in population base; 39.2%) Specific Runoff Load in Dry Weather (t/km²/year)	Runoff in Storm E.1 Load (t/km²/year)	E.1 Storm Rate[a] (%)	E.4 Load (t/km²/year)	E.4 Storm Rate (%)	Term II (sewered rate in population base; 82.4%) Specific Runoff Load in Dry Weather (t/km²/year)	Runoff in Storm E.1 Load (t/km²/year)	E.1 Storm Rate[a] (%)	E.2 Load (t/km²/year)	E.2 Storm Rate[a] (%)
SS	1980	2041.0	25.13	1081.19	97.7	1090.92	97.7	5.01	7.43	59.7	7.96	61.4
	1984	1170.5		488.87	95.1	473.19	95.0		1.21	19.4	1.84	26.9
	1986	1473.0		807.88	97.0	817.26	97.0		7.50	60.0	6.81	57.6
T-COD	1980	2041.0	37.40	329.58	89.8	426.52	91.9	4.96	6.24	55.7	4.68	48.5
	1984	1170.5		168.57	81.8	219.52	85.4		1.50	23.2	1.46	22.7
	1986	1473.0		234.94	86.3	300.42	88.9		5.77	53.8	3.65	42.3
S-COD	1980	2041.0	29.22	9.60	24.7	25.31	46.4	4.12	4.62	52.9	3.23	44.0
	1984	1170.5		3.83	11.6	12.45	29.9		1.36	24.8	1.23	23.0
	1986	1473.0		7.53	20.5	18.60	38.9		4.03	49.5	2.40	36.8

[a] Storm rate = (runoff load in storm)/(runoff load in dry weather + in storm).

flash, and in (2) the Runoff-peak. Therefore, the characteristics of pollutant runoff are examined in these two phases. Each event has such a difference in precipitation, runoff load, and runoff period that it makes it impossible to compare one to another. Standardizing the discharge and the pollutant load of each event with the next equation, the characteristics of pollutant runoff were examined with the Runoff Density Function q(t), l(t) by using ΣQ_{net} and ΣL_{net} shown in Equation 1.

$$t = T / T_r \tag{5}$$

$$q(t) = Q_{net}(T) / \Sigma Q_{net} \tag{6}$$

$$l(t) = L_{net}(T) / \Sigma L_{net} \tag{7}$$

where

T_r = the period since the discharge started to increase until it returned to the origin (sec)

T = time that passed since beginning to react to storm (sec)

$Q_{net}(T)$ = discharge at time T(m³/sec)

$L_{net}(T)$ = load at time T (g/sec)

$\Sigma Q_{net}, \Sigma L_{net}$ = same as in Equation 1 (m³), (g)

7.5.1 Estimation of First-Flash

The phenomenon of first-flash of pollutant runoff during a storm event has not yet been quantified. It was examined qualitatively because of the difficulty involved in quantification. The runoff load during the first-flash is defined with a standardized datum as the integrated runoff load from the beginning of its reaction to a storm up to the maximum point on the first peak. Figure 6 shows its mean value for each year for each pollutant. As Figure 6 shows, the runoff load in the first-flash has a tendency to increase yearly. It is remarkable that these runoff loads are increasing despite progress with a rush of sewering from 1980 to 1989. This is caused by patterns in the basin's drainage system, such as the layout of sewers, considering little urbanization of the basin, i.e., little change in land use, or little expansion of impervious area, etc. Through the drainage system the pollutant load that accumulated on the impervious surface in the urban area, such as on pavement or on the roofs of buildings, runs off quickly during a storm.

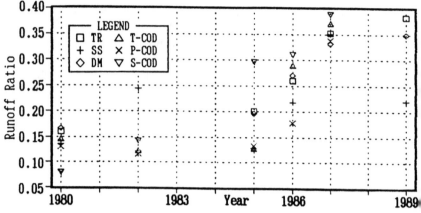

Figure 6.   Transition of runoff ratio at first-flash during storm event.

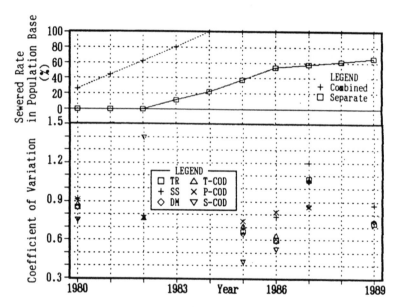

Figure 7.   Transition of coefficient of variation during storm event.

It is thought that the load from nonpoint sources would make up a large part of this runoff load.

7.5.2 Estimation of Runoff Peak

Though the runoff load during a storm event is intensively concentrated around its runoff peak, its characteristics have not been sufficiently estimated to date. Here, the characteristics are examined with the progress of sewering from

1980 to 1989. The fluctuation diagram for pollutant runoff, which formed the runoff peak in response to storm, was developed from the result observed in each event. In order to estimate the extent of these fluctuations as characteristics of the runoff peak, the coefficient of variation (CV) was calculated.

Figure 7 shows the annual mean CV value and the sewered rate in the basin. The CV value on S-COD in 1982 is larger because of the influence of a peculiarly large CV value during the event T-6. Except for this value, the CV value on any pollutant decreased slightly from 1980 to 1985 corresponding to the construction of combined sewers; and increased after 1985 when the combined system already achieved 100% coverage. It is supposed that this depends on both the effect of runoff load reduction by combined sewers and the increase in the direct runoff from nonpoint sources through separate sewers.

As a result, separate sewer systems, instead of combined sewer systems, are being pushed forward today; nevertheless they are not always the best policy, depending on the amount of pollutant runoff during a storm event.

7.6 CONCLUSION

This paper aimed to examine the characteristics of pollutant runoff during a storm event at the point of the receiving water. It attempted:

1. to quantify the effect of sewering in an urban river basin
2. to clear the pattern of pollutant runoff including the first-flash
3. to evaluate the pollutant runoff from nonpoint sources

As a result, the following facts were proved:

1. The reduction effect of sewering is seen on solid pollutants, such as SS and T-COD, both in the dry-weather period and during small storm events, but it is seldom seen on dissolved pollutants or in a large storm.
2. It is presumed that the weight of runoff load from nonpoint sources, particularly during a storm event, is becoming larger in the area of water for public use.
3. Comparing the results of simulation of runoff load through a year using models on a storm event, a large difference can be seen between a flood year and a drought year.
4. Both increase in the runoff load during the first-flash and the sharpening of the runoff peak continue with urbanization in the river basin.

This means that it is extremely significant for the conservation of the environment of water, in particular, to control runoff load from nonpoint sources caused by storms when sewer systems achieve 100% coverage. For this purpose, first, it would be necessary to observe the water quality in an urban river, like the one in this study during a storm event, and second, to develop a monitoring system to do this.

ACKNOWLEDGMENTS

The authors would like to thank Yasunori Nishimoto, a research associate at Ritsumeikan University, and Akikazu Adachi, a graduate student at Tokyo Metropolitan University, for their help in the investigations and for the computer work done to carry out this project.

REFERENCE

1. Ebise, S. "Regression models for Estimation of Storm Runoff Loading," in *Report of the National Institute for Environmental Studies,* Vol. 50. The National Institute for Environmental Studies, Tsukuba, Japan (1984), pp. 59–88.

8 THE TREATABILITY OF URBAN STORMWATER TOXICANTS

8.1 INTRODUCTION

Urban stormwater runoff has been identified as a major contributor to the degradation of many urban streams and rivers.[1-4] Organic and metallic toxicants are expected to be responsible for much of these detrimental effects, and have been found in urban storm-induced discharges during many previous studies.[5-8]

All U.S. cities having populations greater than 100,000 (which total about 15,000 mi^2)[9] will be required to participate in the EPA's stormwater permit program.[10] Based on the Nationwide Urban Runoff Program (NURP) monitoring of the toxicant discharges from 28 cities, it is concluded that urban areas are responsible for substantial toxicant discharges.[5] Although the NURP data was collected mostly from residential areas, with some commercial areas represented, more recent information indicates that industrial stormwater discharges can have many times the concentrations of the toxicants as the areas represented in the NURP data.[11] In addition, dry-weather base flows in separate storm drains may be contaminated by nonstormwater discharges, e.g., industrial waste cross-connections, which can significantly increase the estimated loadings.[12] The EPA sponsored research summarized in this paper was conducted to obtain much needed information concerning the sources and potential control of these stormwater toxicants.

8.2 FIRST PHASE

8.2.1 Introduction

The first phase included the collection and analysis of about 150 urban stormwater runoff and combined sewer overflow (SCSO) samples from a variety

0-87371-805-4/93/$0.00+$.50
© 1993 by Lewis Publishers, Inc.

of source areas and under different rain conditions. A number of the combined sewer overflow (CSO) and detention pond samples were also evaluated. This sampling effort was significantly greater than had been attempted previously for toxic pollutants in stormwater.

Samples were analyzed for organic pollutants using two gas chromatographs (GC), one with a mass selective detector (GC/MSD) and another with an electron capture detector (GC/ECD). Metal concentrations were determined using a graphite-furnace-equipped atomic absorption spectrophotometer (GFAA). All samples were further analyzed for particle-size distributions (from about 1 to 100 μm) and for toxicity using the Microtox® (Microbics, Inc.) toxicity screening technique. All samples were also filtered to determine the liquid/solid partition coefficients of the pollutants and the relative toxicities of the filterable and nonfilterable portions of the samples. Overall, in the first phase of this project about 300 sample components (filterable and total portions of 150 samples) were analyzed to determine toxicant concentrations in sheetflows and other SCSO.

8.2.2 Results

This section summarizes the first project phase results.[8] Most pH values were in a narrow range of 7.0 to 8.5, and the suspended solids concentrations were generally less than 100 mg/L. Particle-size ranges were usually narrow for any one sample, but the ranges for all samples from a single source area category were substantially greater.

Out of more than 35 toxic organic pollutants analyzed, 13 organics were detected in >10% of all samples. The greatest detection frequencies were for 1,3-dichlorobenzene and fluoranthene, which were each detected in 23% of the samples. The organics most frequently found in these samples, polycyclic aromatic hydrocarbons (PAH), especially fluoranthenes and pyrenes, were similar to the organics most frequently detected at outfalls in prior studies.[5]

The toxic organic pollutants had the greatest frequencies of detection in roof runoff, urban creeks, and CSO samples. Vehicle service areas and parking areas had several of the observed maximum organic compound concentrations. Most of the organics were associated with nonfilterable sample portions.

In contrast to the organics, the heavy metals analyzed were detected in almost all samples, including those filtered. Roof runoff had the highest concentrations of zinc, probably from galvanized roof drainage components. Parking areas had the highest nickel concentrations, while vehicle service areas had the highest concentrations of cadmium and lead. Urban creek samples had the highest copper concentrations, probably from illicit connections or nonstormwater entries.

During the first phase, about 20 SCSO samples were analyzed using a variety of laboratory bioassay tests including the Microtox® screening procedure. The results indicated that the Microtox® procedure gave similar toxicity rankings as most of the other bioassays. The Microtox® procedure was not used to determine if the samples were toxic, or to predict the toxic effects of stormwater runoff on receiving waters. This screening procedure was instead used to identify the source areas having the most toxic runoff and to determine the toxicity differences between the filterable and nonfilterable sample components. In-stream taxonomic investigations are needed to further interpret actual toxicity problems. Laboratory bioassay tests can be useful to determine the major sources of toxicants and to investigate toxicity reduction through treatment but they are not a substitute for actual in-stream investigations of receiving-water effects.

About 15% of the unfiltered samples were considered highly toxic using the Microtox® screening procedure. The remaining samples were approximately evenly split between being moderately toxic and not toxic. The Microtox® screening tests found that the greatest percentage of samples considered the most toxic were from CSO, followed by samples obtained from parking and industrial storage areas. Runoff from paved areas had relatively low suspended solids concentrations and turbidities, especially compared to samples obtained from unpaved areas.

Data evaluations indicated that variations in Microtox® toxicities and organic toxicant concentrations may be greater for different rains than for different source areas. As an example, high concentrations of PAH were more associated with long antecedent-dry periods than with other rain parameters or source area or land-use sampling locations.

A literature review conducted found that many processes will affect the potential transport and fate mechanisms of these pollutants. Sedimentation in the receiving water is the most common fate mechanism because many of the pollutants investigated are associated with particulate matter. Exceptions include zinc and 1,3-dichlorobenzene, which are mostly associated with the filterable sample portions. Particulate removal can occur in many SCSO control facilities including (but not limited to) catch basins, swirl concentrators, screens, filters, drainage systems, and detention ponds. These control facilities (with the possible exception of drainage systems) allow removal of the accumulated polluted sediment for final disposal in an appropriate manner. Uncontrolled sedimentation will occur in receiving waters, such as lakes, reservoirs, or large rivers. In these cases, the wide dispersal of the contaminated sediment is difficult to remove and can cause significant detrimental effects. Biological or chemical degradation of the sediment toxicants may occur but in the expected anaerobic environments it is quite slow for many of the pollutants. Degradation by photochemical reaction and volatilization (evaporation) of the soluble pollutants may also occur, especially when these pollutants are near the surface of aerated waters. Increased turbulence and aeration encourages the processes, which in turn may significantly reduce toxicant concentrations. In contrast, quiescent waters would encourage sedimentation that would also reduce toxicant concentration. Metal precipitation and sorption of pollutants onto suspended solids increases the sedimentation and/or flotation potential of the pollutants and also encourages more efficient bonding of the pollutants to soils particles, preventing their leaching to surrounding waters.

8.3 SECOND PHASE

8.3.1 Introduction

In the second project phase toxicants treatability evaluations of a variety of bench-scale treatment processes were conducted. Later project phases will examine the treatability of SCSO toxicants in greater detail and address the modification of existing treatment processes.

8.3.2 Sampling Effort and Experimental Error

The relative importance of different source areas, e.g., roofs, streets, parking areas, etc., in contributing toxicants was determined from the examination of 150 source area samples during the first project phase.[8] These samples were collected from the most potentially toxic pollutant source areas in residential, commercial, and industrial land uses. The areas that received the greatest emphasis (or most sampling) during both project phases were parking and storage areas in industrial and commercial lands. These areas had been noted in previous studies to have the largest potential of discharging toxicants.[11] First-phase sheet flow samples were collected during five rains in Birmingham, Alabama. Replicate samples taken from many of the same source areas, but during different rains, enabled differences due to rain conditions vs. site locations to be statistically evaluated.

The second phase included the intensive analyses of 12 samples. Table 1 lists these samples, their sampling dates, and their source area categories. These sampled rains represent practically all of the rains that occurred in the Birmingham area during the field portion of the second project phase (July to November, 1990). Table 1 also includes information concerning the toxicities of the samples before they were subjected to treatment. Independent replicates were used to determine the measurement errors associated with the Microtox® procedure. The total number of Microtox® analyses that were conducted for all of the treatability tests for each sample are also noted, as are the means, standard deviations, and coefficients of variation of the replicate toxicity values.

Initial toxicity values (before treatability tests) were plotted on normal-probability graphs to indicate their probability distributions. Almost all of the samples had initial toxicity values that were shown to be normally distributed. Therefore, the coefficient of variation values shown on Table 1 can be used as an indication of the confidence intervals of the Microtox® measurements. The coefficients of variation ranged from 2.3 to 9.8%, with an average value of 5.1%.

Table 1. Sample Descriptions

Automobile Service Area Samples

Sample	Date	Toxicity (% light reduction)	Number of Analyses	Standard Deviation	Relative Standard Deviation (%)
B	7/10/90	78	28	7.6	9.8
C	7/21/90	34	42	2.9	8.5
E	8/19/90	43	74	1.3	3.0
H	10/17/90	50	88	1.5	3.0

Industrial Loading and Parking Area Sample

Sample	Date	Toxicity (% light reduction)	Number of Analyses	Standard Deviation	Relative Standard Deviation (%)
D	8/2/90	67	74	2.1	3.1
F	9/12/90	31	88	1.5	4.9
G	10/3/90	53	88	3.0	5.7
I	10/24/90	55	89	1.9	3.4
J	11/5/90	49	89	1.1	2.3
K	11/9/90	28	89	2.2	8.1

Automobile Salvage Yard Samples

Sample	Date	Toxicity (% light reduction)	Number of Analyses	Standard Deviation	Relative Standard Deviation (%)
L	11/28/90	26	89	1.4	5.5
M	12/3/90	54	89	1.8	3.4

Overview

		Toxicity (% light reduction)	Number of Analyses	Standard Deviation	Relative Standard Deviation (%)
minimum:		26		1.1	2.3
maximum:		78		7.6	9.8
mean:		47		2.4	5.1
st. dev.:		16			
total:			927		

Therefore, the 95% confidence interval (two times the relative standard deviation values include 95.4% of the values, if normally distributed) for the Microtox® procedure ranged between 5 and 20% of the mean values. These confidence intervals are quite narrow for a bioassay test and indicate the good repeatability of the Microtox® procedure. One of the important features of the Microtox® test is the use of a very large number of organisms (about one million) for each analysis, reducing erratic test responses that may be caused by unusual individual

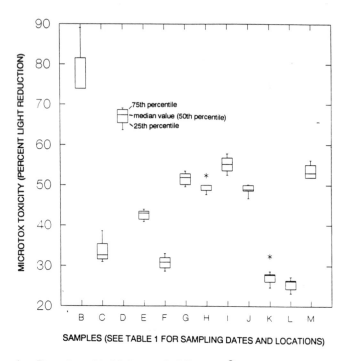

Figure 1.   Box plot of initial sample Microtox® toxicities.

organisms. In all cases, statistical tests were performed on the test results to indicate the significance of the different treatability tests.

Figure 1 contains box plots of the initial toxicity values.[13] These indicate the spread of toxicity values that were represented by the samples. Two samples (B and D) were found to be highly toxic, while the remainder were moderately toxic.

8.3.3 Sampling Procedures

Sheet flow samples were collected using manual grab procedures. For deep sheet flows, samples were collected directly into the sample bottles, or dipped using glass beakers. For shallow sheet flows, hand operated pumps created a vacuum in the sample bottle which then drew the sample directly into the container through a Teflon™ tube. About 10 to 20 L of each sample were collected for the treatability analyses. The samples were all obtained from the Birmingham, Alabama, area.

8.3.4 Toxicity Screening Tests

A number of previous studies have found high concentrations of toxic pollutants in stormwater samples. Some urban stormwater runoff studies attempted to use conventional 96-h fathead minnow fish bioassay toxicity tests.[14]

Very few fish died during the tests; however, *in situ* taxonomic studies of urban stormwater runoff receiving waters (including the same stream used for the negative fish bioassay tests) found significant evidence of toxic effects from the long-term exposure to these pollutants.[12] Recent bioassay tests have used more sensitive organisms and have detected significant SCSO toxicities.[15-18]

The objective of these toxicity analyses was to obtain relative toxicity measurements from a large number of subsamples from different stages of bench-scale treatment (treatability) tests. These tests were not used to determine the absolute toxicities of the samples but only to examine the toxicity differences between the sample partitions from different treatment tests. To evaluate the different treatment options, it was necessary to use a rapid screening method that only used small sample volumes. The Microtox® toxicity testing procedure uses marine phosphorescent algea to indicate relative toxicities of samples. As noted earlier, the first project-phase comparison of a variety of unconventional and conventional laboratory bioassay tests indicated that the Microtox® procedure gave toxicity rankings similar to the other tests.

Toxicity, as determined by the Microtox® procedure, was expressed as three values, I_{10} (the percent light decrease after 10 min of exposure), I_{35} (the percent light decrease after 35 min of exposure), and the EC_{50} (the sample dilution corresponding to a 50% light decrease after a 35 min exposure). Only samples that have I_{35} values >50 were further tested to determine the EC_{50} values. Higher values of I_{10} and I_{35}, and lower fractions of EC_{50}, correspond to greater relative toxicities.

Microbics[R] suggests that light decrease values greater than 60% correspond to "highly" toxic samples, light decrease values between 20 and 60% correspond to "moderately" toxic samples, and light decrease values less than 20% correspond to "nontoxic" samples.

During the first project phase, a number of special tests were conducted that examined problems associated with sample storage time, preservation, and sample containers. Teflon™ and glass were exclusively used to reduce the toxic effects of the containers; samples were stored at 3 to 5 °C and examined within 24 h of sample collection; and the required osmotic adjustments were not made until immediately before sample analysis in order to minimize chemical reactions between the NaCl and the toxicants.

8.3.5 Solids Physical Characterization

Introduction

Most SCSO physical treatment device removal efficiencies significantly relate to solids particle size and/or settling velocity distributions.[19] Wet detention/storage tanks and ponds, grass filters, street cleaning, microscreening, filtration, and swirl concentrators are some of the pollution control devices that require a knowledge of particle size and/or settling characteristics. Additionally, the fates of many toxic pollutants in receiving waters are also very sensitive to the physical characteristics of particles. Without knowing the particle size and settling velocity distributions of the SCSO it is impossible to correctly design many of these physical treatment devices.

Particle-Size Analyses

During the different treatment phases, a laser particle counter (SPC-510 from Spectrex Corp.) was used to analyze particle-size distributions of all of the samples. This instrument produces particle size distribution plots for particle-sizes ranging from 0.5 µm to more than 100 µm.

Settling Velocity and Turbidity Analyses

In addition, settling column tests were concurrently conducted to determine the settling velocity and also specific gravity distributions, thereby allowing the calculation of velocity distributions for other particle-size distribution data sets. Nephelometric turbidity analyses were conducted for all subsamples during the treatability tests.[20] Gravimetric solids analyses were conducted on all settling column subsamples to calculate settling velocities and specific gravity.[21]

Treatability Tests

The second phase included tests to examine the treatability of toxicants in source area samples. As previously noted, all samples were relatively toxic. This allowed a wide range of laboratory partitioning and treatability analyses to be conducted without detection limit problems. The following treatability tests were examined in this study:

- settling column (a 1.5 in. x 30 in. Teflon™ column)
- floatation (a series of eight 100-mL narrow-neck (volumetric) glass flasks)
- screening and filtering (a series of eleven stainless steel sieves, from 20 to 106 µm, plus a 0.45 µm membrane filter)
- photodegradation (a 2-L glass beaker with a 60-W broad-band incandescent light placed 6 in. above the water, stirred with a magnetic stirrer with water temperature and evaporation rate also monitored)
- aeration (the same beaker arrangement as above, without the light, but with filtered compressed air keeping the test solution supersaturated and well mixed)
- photodegradation and aeration combined (the same beaker arrangement as above, with compressed air, light, and stirrer)
- chemical coagulation/flocculation (standard jar tests using alum and 800-mL samples)
- undisturbed control sample (a sealed, wrapped, and covered glass jar at room temperature)

Because of the difficulty of obtaining large sample volumes from many of the source areas that were to be examined, these bench-scale tests were all designed to use small sample volumes.

Each test (except for filtration and chemical coagulation) was conducted over a period of time of up to 3 d. Subsamples were typically obtained for toxicity analyses at 0, 1, 2, 3, 6, 12, 24, 48, and 72 h during the tests. In addition, settling column samples were obtained several times within the first hour, including at 1, 3, 5, 10, 15, 25, and 40 min. The chemical coagulation tests were conducted using several concentrations of alum in a standard jar test. In addition to the Microtox® toxicity tests, most samples were analyzed for turbidity and particle sizes. All settling column samples used gravimetric suspended solids analyses to enable calculations of settling velocity to be made.

Future project phases will include pilot- and full-scale tests of the most promising control and treatment practices. Especially important in the future project phases will be the testing of modifications to conventional SCSO treatment devices and the design and testing of combination treatment systems suitable for small source areas (such as pavement at automobile service facilities, especially gasoline service stations).

Data Observations

The Microtox® procedure allowed toxicity screening tests to be conducted on each sample partition during the treatment tests. This efficient procedure enabled more than 900 toxicity tests to be made. Turbidity and particle-size distribution tests were also made on all samples.

Figures 2 to 10 are graphical data plots of the toxicity reductions observed during each treatment procedure examined, including the control measurements. Each figure contains three graphs. One contains the treatment responses for the automobile service facility (samples B, C, E, and H), another for the industrial loading and parking area (samples D, F, G, I, J, and K), and the last one for the automobile salvage yard (samples L and M).

Even though the data are separated into these three groups, very few consistent differences are noted in the way the samples responded to various treatments. As expected, there are greater apparent differences between the treatment methods than between the sample groupings. Statistical tests that will be conducted during the current project phase will examine these groupings in detail.

Tables 2 to 4 summarize results from the nonparametric Wilcoxon signed ranks test (using SYSTAT: The System for Statistics, Version 5, SYSTAT, Inc., Evanston, Illinois) for different treatment combinations. This statistical test indicates the two-sided probabilities that the sample groups are the same. A probability of 0.05 or less is used to indicate significant differences in the data sets. As an example, Table 3 indicates that for sample D the undisturbed control

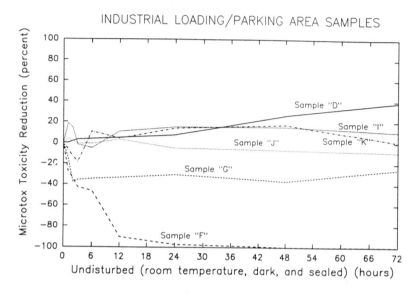

Figure 2a.

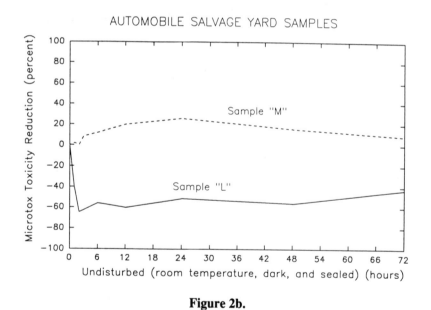

Figure 2b.

Figure 2.   Undisturbed sample toxicity trends.

AUTOMOBILE SERVICE AREA SAMPLES

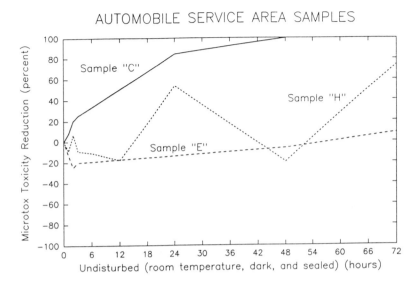

Figure 2c.

INDUSTRIAL LOADING/PARKING AREAS

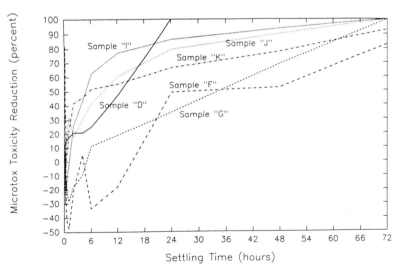

Figure 3a.

Figure 3.   Settling column treatability test toxicity trends.

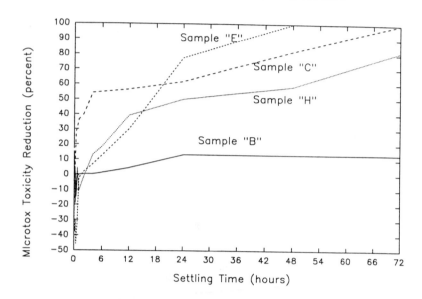

Figure 3b.

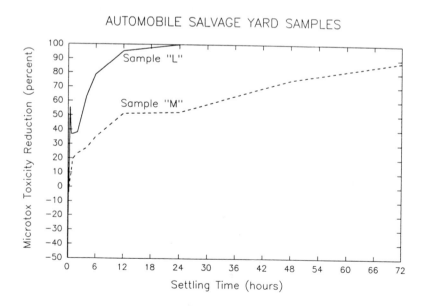

Figure 3c.

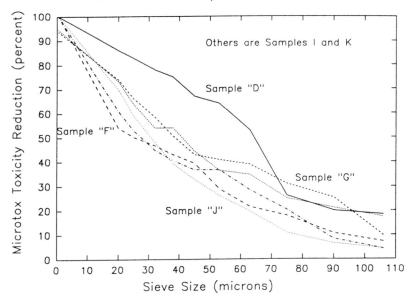

Figure 4a.

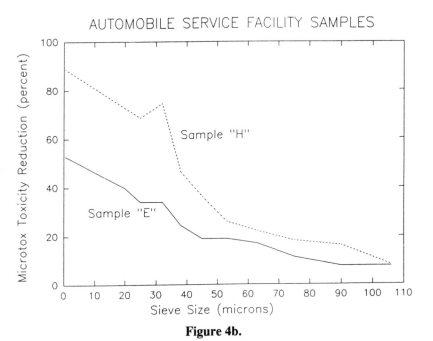

Figure 4b.

Figure 4.   Sieve treatability test toxicity trends.

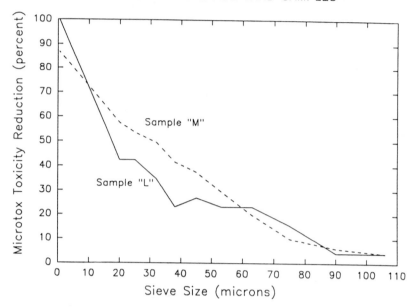

Figure 4c.

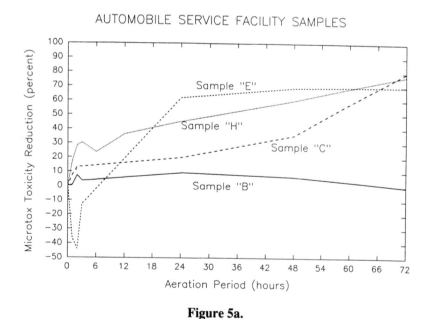

Figure 5a.

Figure 5.   Aeration treatability test toxicity trends.

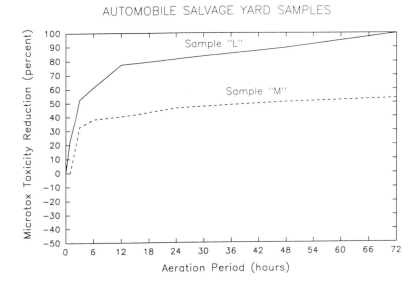

Figure 5b.

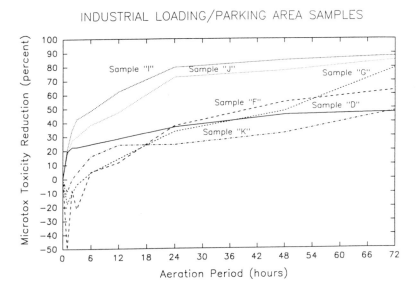

Figure 5c.

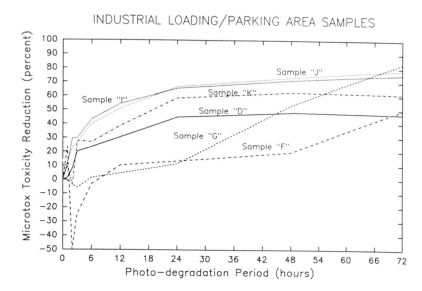

Figure 6a.

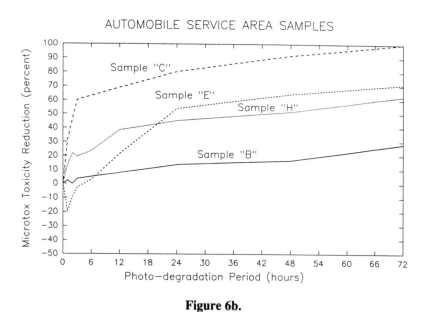

Figure 6b.

Figure 6. Photodegradation treatability test toxicity trends.

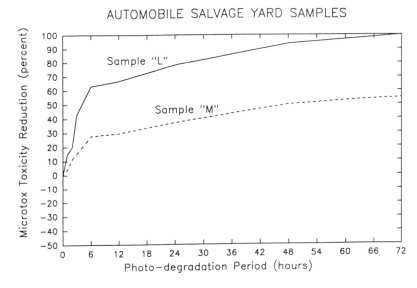

Figure 6c.

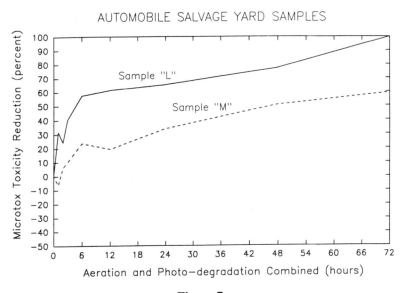

Figure 7a.

Figure 7.   Aeration and photodegradation combined treatability test toxicity trends.

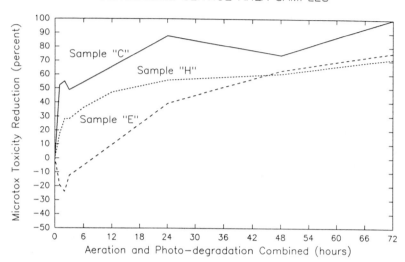

Figure 7b.

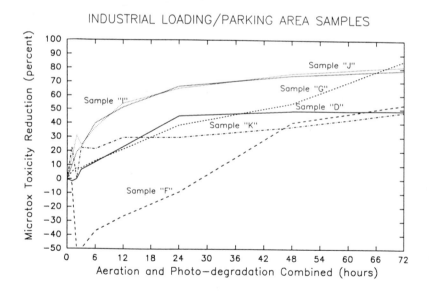

Figure 7c.

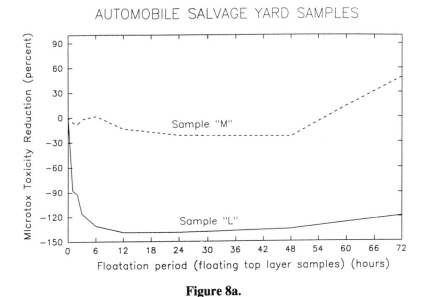

Figure 8a.

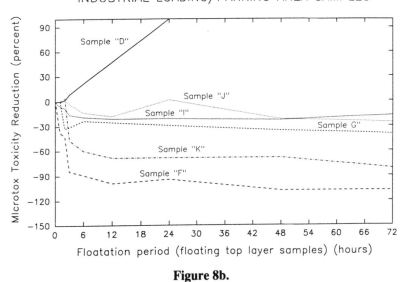

Figure 8b.

Figure 8.   Flotation treatability test toxicity trends (top layer sample).

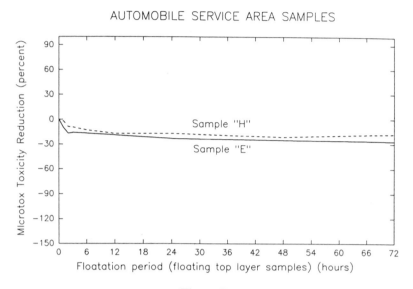

Figure 8c.

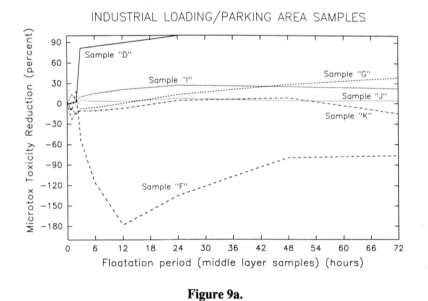

Figure 9a.

Figure 9.   Flotation treatability test toxicity trends (middle layer samples).

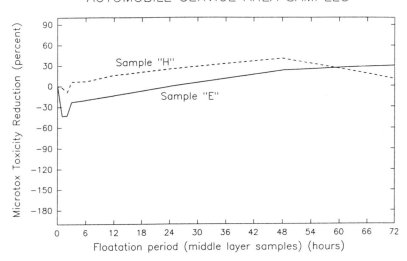

Figure 9b.

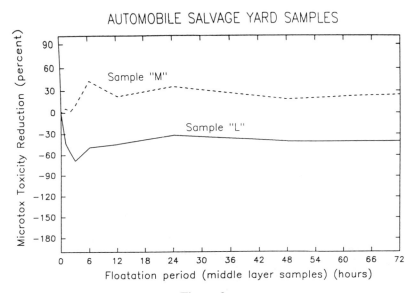

Figure 9c.

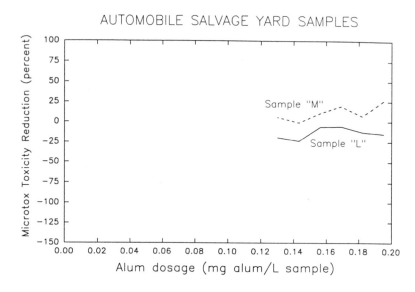

Figure 10a.

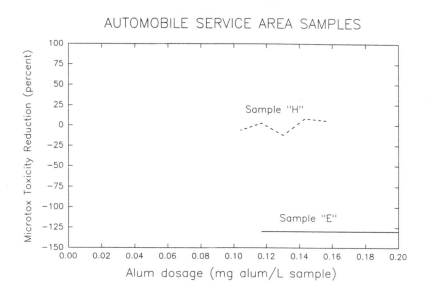

Figure 10b.

Figure 10.   Alum addition treatability test toxicity trends.

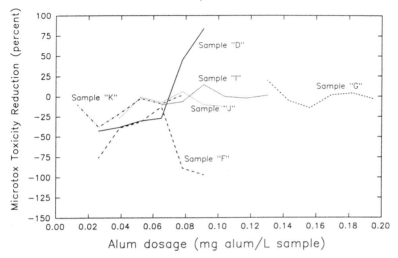

Figure 10c.

sample was significantly different (with probabilities of 0.02) compared to all of the treatment tests.

The aeration test provided the most samples that had significant probabilities of being different from the control condition. Settling, photodegradation, and aeration and photodegradation combined, were similar in providing the next greatest number of samples that had significant probabilities of being different from the control condition. The flotation test had many samples that had significant differences in toxicity of the top floating layer compared to the control sample. However, the more important contrast between the middle sample layers (below the top floating layer) and the control sample, which would indicate a reduction in toxicity of post-treated water, had very few samples that were significantly different from the control sample.

The absolute magnitudes of toxicity reductions must also be considered. As an example, it may be significant, but unimportant, if a treatment test provided many (and therefore consistent) samples having statistically significant differences compared to the control sample, if the actual toxicity reductions were very small.

Conclusions

As shown on Figures 2 to 10 good separation of toxicant responses were found during many of the treatment tests. The highest toxicant removals were obtained

TABLE 2. Two-sided Probabilities Comparing Different Treatment Tests for Automobile Service Area Samples[a]

Automobile Service Area Samples:	B	C	E	H
Undisturbed versus:				
settling	n/a	0.25	0.02	0.41
aeration	n/a	0.31	0.25	0.07
photodegradation	n/a	0.12	0.06	0.16
aeration and photodegradation	n/a	0.35	0.24	0.06
flotation – top layer	n/a	n/a	0.74	0.02
flotation – middle layer	n/a	n/a	0.31	0.87
Aeration and Photodegradation:				
aeration vs. photodegradation	0.23	0.02	0.49	0.08
aeration vs. aeration and photo.	n/a	0.03	0.99	0.14
photo. vs. aeration and photo.	n/a	0.25	0.14	0.02
Flotation:				
top layer vs. middle layer	n/a	n/a	0.49	0.01
Settling versus:				
aeration	0.46	0.02	0.02	0.45
photodegradation	0.12	0.25	0.02	0.79
aeration and photodegradation	n/a	0.61	0.02	0.09
flotation – top layer	n/a	n/a	0.02	0.05
flotation – middle layer	n/a	n/a	0.02	0.09
Aeration versus:				
flotation – top layer	n/a	n/a	0.39	0.02
flotation – middle layer	n/a	n/a	0.21	0.02
Photodegradation versus:				
flotation – top layer	n/a	n/a	0.18	0.02
flotation – middle layer	n/a	n/a	0.03	0.02
Aeration and Photodegradation versus:				
flotation – top layer	n/a	n/a	0.49	0.02
flotation – middle layer	n/a	n/a	0.04	0.02

[a]   Probabilities were calculated using the Wilcoxon signed-rank test for paired data sets. Comparisons having probabilities less than, or equal to, 0.05 are considered significantly different.

TABLE 3. Two-sided Probabilities Comparing Different Treatment Tests for Industrial Loading and Parking Area Samples[a]

Industrial Loading and Parking Area Samples:	D	F	G	I	J	K
Undisturbed versus:						
settling	0.02	0.12	0.09	0.07	0.01	0.01
aeration	0.02	0.05	0.06	0.04	0.01	0.01
photodegradation	0.02	0.04	0.03	0.07	0.01	0.01
aeration and photodegradation	0.02	0.05	0.03	0.09	0.01	0.01
floatation – top layer	0.02	0.05	0.13	0.01	0.03	0.21
floatation – middle layer	0.02	0.78	0.02	0.26	0.16	0.17
Aeration and Photodegradation:						
aeration vs. photodegradation	0.21	0.24	0.74	0.01	0.04	0.05
aeration vs. aeration and photo.	0.61	0.18	0.04	0.01	0.11	0.51
photo. vs. aeration and photo.	0.21	0.16	0.25	0.79	0.74	0.12
Flotation:						
top layer vs. middle layer	0.72	0.41	0.05	0.02	0.07	0.12
Settling versus:						
aeration	0.18	0.33	0.61	0.48	0.41	0.02
photodegradation	0.02	0.78	0.61	0.06	0.12	0.02
aeration and photodegrad.	0.03	0.67	0.75	0.05	0.12	0.03
floatation – top layer	0.14	0.05	0.13	0.04	0.01	0.02
floatation – middle layer	0.72	0.09	0.31	0.05	0.02	0.01
Aeration versus:						
flotation – top layer	0.39	0.04	0.09	0.02	0.01	0.09
flotation – middle layer	0.12	0.09	0.18	0.03	0.01	0.01
Photodegradation versus:						
flotation – top layer	0.04	0.04	0.04	0.02	0.01	0.09
flotation – middle layer	0.04	0.05	0.24	0.04	0.01	0.01
Aeration and Photodegrad. versus:						
flotation – top layer	0.03	0.04	0.06	0.02	0.01	0.05
flotation – middle layer	0.18	0.21	0.04	0.04	0.01	0.01

[a] Probabilities were calculated using the Wilcoxon signed-rank test for paired data sets. Comparisons having probabilities less than, or equal to, 0.05 are considered significantly different.

TABLE 4. Two-sided Probabilities Comparing Different Treatment Tests for Automobile Salvage Yard Samples[a]

Automobile Salvage Yard Samples:	L	M
Undisturbed versus:		
settling	0.02	0.02
aeration	0.02	0.03
photodegradation	0.02	0.16
aeration and photodegradation	0.02	0.09
flotation – top layer	0.01	0.09
flotation – middle layer	0.59	0.89
Aeration and Photodegradation:		
aeration vs. photodegradation	0.08	0.01
aeration vs. aeration and photodegradation	0.07	0.08
photodegradation vs. aeration and photodegradation	0.99	0.14
Flotation:		
top layer vs. middle layer	0.02	0.07
Settling versus:		
aeration	0.02	0.12
photodegradation	0.02	0.01
aeration and photodegradation	0.02	0.02
flotation – top layer	0.02	0.02
flotation – middle layer	0.02	0.03
Aeration versus:		
flotation – top layer	0.02	0.01
flotation – middle layer	0.02	0.01
Photodegradation versus:		
flotation – top layer	0.02	0.03
flotation – middle layer	0.02	0.21
Aeration and Photodegradation versus:		
flotation – top layer	0.02	0.01
flotation – middle layer	0.02	0.16

[a] Probabilities were calculated using the Wilcoxon signed-rank test for paired data sets. Comparisons having probabilities less than, or equal to, 0.05 are considered significantly different.

by settling for at least 24 h (providing generally 40 to 90% reductions), screening through at least a 40 μm screen (20 to 70% reductions), and aeration and/or photodegradation for at least 24 h (up to 80% reductions). Increased settling, aeration or photodegradation times, and screening through finer meshes, all resulted in greater toxicity reductions. The floatation tests produced floating sample layers that generally increased in toxicity with time and lower sample layers that generally decreased in toxicity with time, as expected; however, the benefits were quite small (less than 30% reduction). Alum coagulation/flocculation substantially reduced the turbidities of the samples, but the changes in toxicity were highly irregular. These results, in conjunction with results from the first project phase, will enable the modification of treatment device/system designs (for new installations and for retrofitting existing installation) in order to optimize toxicant removals from critical stormwater runoff source areas.

REFERENCES

1. Field, R., and R. Turkeltaub. "Urban runoff receiving water impacts: program overview," *J. Environ. Eng.* 107: 83 (1981).
2. Pitt, R., and M. Bozeman. "Sources of urban runoff pollution and its effects on an urban creek," U.S. EPA Report-600/2-82-090. (December 1982).
3. Pitt, R., and P. Bissonnette. "Bellevue urban runoff program, summary report," U.S. Environmental Protection Agency and the Storm and Surface Water Utility, Bellevue, WA (1984).
4. Field, R., and R. Pitt. "Urban storm-induced discharge impacts: U.S. Environmental Protection Agency research program review," *Water Sci. Technol.* 22: 1 (1990).
5. "Water Planning Division. Results of the nationwide urban runoff program," U.S. EPA Water Planning Division, NTIS no. PB 84-185552 (1983).
6. Hoffman, E.J., G.L. Mills, J.S. Latimer and J.G. Quinn. "Urban runoff as a source of polycyclic aromatic hydrocarbons to coastal waters," *Environ. Sci. Technol.* 18: 580 (1984).
7. Fam, S., M.K. Stenstrom and G. Silverman. "Hydrocarbons in urban runoff," *J. Environ. Eng.* 113: 1032 (1987).
8. Pitt, R., and P. Barron. "Assessment of urban and industrial stormwater runoff toxicity and the evaluation/development of treatment for runoff toxicity abatement – Phase 1," Draft Report, U.S. EPA Contract #68-C9-0033, Edison, NJ (1990).
9. "Statistical Abstract of the United States, 1980," U.S. Department of Commerce, Bureau of the Census, 101st ed., Washington, D.C. (1980).
10. "Final rule, national pollutant discharge elimination system permit application regulations for storm water discharges," U.S. EPA Report-40 CFR Parts 122–124 (November 16, 1990).

11. Pitt, R., and J. McLean. "Toronto area watershed management strategy study: Humber River pilot watershed project," Ontario Ministry of the Environment, Toronto, Ontario (1986).

12. Pitt, R., M. Lalor, M. Miller and G. Driscoll. "Assessment of non-storm-water discharges into separate storm drainage networks – Phase 1: development of methodology for a manual of practice," U.S. EPA Draft Report, Contract #68-C9-0033, Office of Research and Development, Edison, NJ (1990).

13. Pitt, R. "Demonstration of nonpoint pollution abatement through improved street cleaning practices," U.S. EPA Report-600/2-79-161, Cincinnati, OH (1979).

14. Tukey, J.W. *Exploratory Data Analysis,* (Reading, MA: Addison-Wesley, 1977).

15. Spiegel, S.J., E.C. Tifft, C.B. Murphy and R.R. Ott. "Evaluation of urban runoff and combined sewer overflow mutagenicity," U.S. EPA Report-600/2-84-116, Cincinnati, OH (1984).

16. Mount, D.I., A.E. Steen and T.J. Norberg-King. "Validity of effluent and ambient toxicity for predicting biological impact on Five Mile Creek, Birmingham, Alabama," U.S. EPA Report-600/8-85/015, Duluth, MN (1985).

17. Mount, D.I., T.J. Norberg-King and A. E. Steen. "Validity of effluent and ambient toxicity for predicting biological impact, Naugatuck River, Waterbury, Connecticut," U.S. EPA Report-600/8-86/005, Duluth, MN (1986).

18. Norberg-King, T.J., D.I. Mount, J.R. Amato and J.E. Taraldsen. "Application of the water quality based approach on a regional scale: toxicity testing and identification of toxicants in effluents from the San Francisco Bay Region," U.S. EPA Technical Report 01-88, Duluth, MN (1988).

19. Dalrymple, R.J., S.L. Hodd and D.C. Morin. "Physical and settling characteristics of particulates in storm and sanitary wastewaters," U.S. EPA Report-670/2-75-011, Cincinnati, OH (1975).

20. Method 180.1, in "Methods for chemical analysis of water and wastes," U.S. EPA Report-600/4-79-020, Cincinnati, OH (1983).

21. Method 160, in "Methods for chemical analysis of water and wastes," U.S. EPA Report 600/4-79-020, Cincinnati, OH (1983).

9 HYDROLOGIC AND WATER QUALITY COMPARISONS OF RUNOFF FROM POROUS AND CONVENTIONAL PAVEMENTS

9.1 INTRODUCTION

9.1.1 Urban Runoff Characteristics

The impacts of urban development on local water resources generally include stormwater runoff peaks and volumes of greater magnitude than in the predeveloped state, often occurring in association with a degradation of receiving-water quality. The major reason for these impacts is the development of impervious areas, such as roofs, streets, and parking lots, which reduce the infiltration capacity of urban watersheds and produce a corresponding increase in runoff rates and volumes.

Stormwater management generally consists of collecting and transporting overland runoff in a conveyance system of storm sewers or channels that are tributary to a nearby stream or lake. Although local flooding problems may be solved by this method, the shorter time of concentration and higher peak flows which are generated may create more severe flood problems downstream. The increase in flow velocities in the improved channels creates a high erosion and scour potential, thus exacerbating the problem of pollutant transport to receiving bodies of water.

Stormwater flows transport contaminants, which accumulate on the watershed during dry weather; however, the total mass transported is a function of the contaminant accumulation rate, the number of preceding dry days, the intensity of the rainfall, the velocity and volume of surface flow, and other physical properties of the catchment. Impervious areas generally have limited assimilative properties and, in some cases, tend to yield contaminants that are not

0-87371-805-4/93/$0.00+$.50
© 1993 by Lewis Publishers, Inc.

amenable to control and removal using standard maintenance procedures. Heavy metals, oils, and other hydrocarbons from automobiles and machinery, suspended solids from dust and dirt accumulation, and airborne pollutants washed out during precipitation events are typical contaminants present in urban stormwater runoff.

To protect the water quality of receiving waters, stormwater runoff is generally routed either overland or through storm drains to a detention facility which is an effective removal mechanism for suspended constituents by settling. Other stormwater management options include:

1. groundwater recharge by filtration through sand or other media (and prevention of clogging by scraping the filter beds periodically)
2. maintenance of natural drainage patterns and limiting the amount of impervious cover to prevent excessive erosion and to allow deposition of suspended stormwater constituents
3. runoff-flow energy dissipation efforts to reduce erosion potential
4. surface drainage channel design criteria
5. general development restrictions
6. treatment charges based on the contaminant load in the runoff

9.1.2 Project Objectives

Three specific objectives of the project considered in this paper are as follows:

1. to determine the stormwater hydrologic and water quality characteristics of porous pavements and to compare these characteristics to nonporous pavements
2. to determine the relative capability of porous and nonporous pavements to assimilate or reduce typical pollutants in urban runoff through storage and percolation
3. to compare the performance of these pavements under a range of storm conditions

9.2 STUDY APPROACH

9.2.1 Overview

An extensive monitoring program[1] was initiated to document the hydraulic and pollutant transport characteristics of several porous and conventional pavement facilities located in Austin, Texas. Five parking lots, representing a variety of pavement surfaces, were selected:

A. Porous Surfaces

1. porous asphalt-concrete
2. lattice block
3. gravel trench at the edge of a conventional asphalt- concrete lot

B. Nonporous Surfaces

4. conventional asphalt-concrete
5. conventional concrete

The City of Austin is located on the Colorado River of Texas. Elevations within the city vary from 400 ft to 900 ft (125 m to 275 m) above mean sea level. The mean annual temperature is 68.1°F (20.1° C); winters are mild and summers are hot. Average annual rainfall in the Austin area is 32.49 in. (82.5 cm).

9.2.2 Porous Pavements

A porous pavement system is an innovative solution to the problem of controlling water drainage and detention from parking and other low-traffic areas. A schematic cross section of a typical asphalt-concrete porous pavement facility is presented in Figure 1. This type of pavement can utilize the natural infiltration capacity of the soil to absorb rainfall and local runoff after accumulation in a porous base consisting of sand or large diameter open graded gravel. If infiltration into the soil is undesirable or not practical, lateral drainage to a sump or channel can be provided.

In regular applications for highway and airport runway construction, a commonly used porous pavement surface has been referred to as plant mix seal coat, open-graded mix, gap-graded mix, popcorn mix, or porous friction course. The resulting paving has a coarse surface texture and a high void ratio resulting in temporary storage of surface water while maintaining the coefficient of friction between a vehicle tire and pavement at values comparable to the coefficient under dry conditions.

Parking lots consisting of concrete lattice blocks with grass planted in the interstices are an aesthetic and practical solution to urban stormwater drainage and detention. A typical cross section through this type of paving is shown in Figure 2. This approach also allows for storage of runoff in the low areas within the blocks, and permanently removes that volume of water from surface discharge. Furthermore, the grass may provide an added benefit in being able to recycle nutrients and other runoff constituents.

Porous pavements can be designed to retain all of the rainfall and runoff with no drainage from the site, to retain sufficient rainfall and runoff to reduce the after-development hydrologic conditions to predevelopment conditions, or to delay runoff from the site, thus attenuating peak discharges and reducing the impact of associated pollutant transport.

9.2.3 Description of Study Sites

The physical characteristics of the five parking lot pavement surfaces selected for monitoring are presented in Table 1. The following text discusses the physical characteristics and sampling procedures for each study lot.

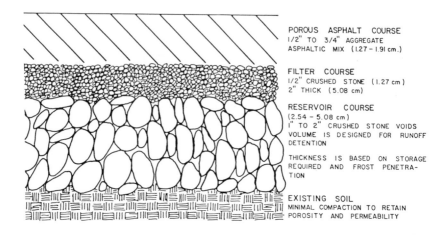

POROUS ASPHALT COURSE
1/2" TO 3/4" AGGREGATE
ASPHALTIC MIX (1.27 – 1.91 cm)

FILTER COURSE
1/2" CRUSHED STONE (1.27 cm)
2" THICK (5.08 cm)

RESERVOIR COURSE
(2.54 – 5.08 cm)
1" TO 2" CRUSHED STONE VOIDS
VOLUME IS DESIGNED FOR RUNOFF
DETENTION

THICKNESS IS BASED ON STORAGE
REQUIRED AND FROST PENETRA-
TION

EXISTING SOIL
MINIMAL COMPACTION TO RETAIN
POROSITY AND PERMEABILITY

Figure 1.   Porous asphalt paving typical section.

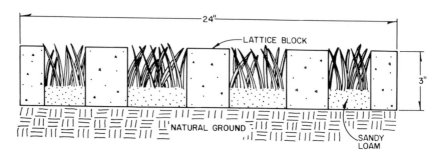

Figure 2.   Typical lattice block section.

Porous Asphalt Lot

The porous asphalt lot consisted of a surface asphalt layer and two layers of stone base course with rocks ranging from 1.5 to 2.5 in. (3.0 to 6.5 cm) in diameter. The lower base course layer ranged in depth from approximately 4 in. (10 cm) on the upslope end to 42 in. (107 cm) on the low end and provided a void space of approximately 35 percent of its gross volume for water retention. The upper layer averaged 2 in. (5 cm) in depth and consisted of a relatively uniform gravel course (filter course) with material ranging from 0.4 to 0.6 in. (1.0 to 1.5

TABLE 1. Physical Characteristics of the Study Lots

Pavement Surface	Location	Area[a] (acres)	Slope (%)	Parking Spaces	Traffic[b] (vehicles per day)
Porous asphalt	City maintenance yard	0.354	(surface) 0.0 (base) 2.6	20	10
Lattice block	Symphony Square	0.143	4.0	14	14
Gravel trench at asphalt lot	Airport rental car storage	1.364	5.0	186	375
Conventional asphalt	Sr. Citizen's Activity Center	0.255	3.9	33	100
Conventional concrete	City Hall Annex	0.369	1.1	35	70

[a] 1 acre = 43,560 ft^2 = 4,047 m^2.
[b] Estimated from the number of parking spaces and field observations of approximate daily occupancy rates.

cm) in diameter. This intermediate layer was selected to provide a stable and uniform surface for the application of the porous asphalt-concrete. The final surface layer consists of a 2.5-in. (6.4-cm) thick porous asphalt-concrete with 5.5 to 6.0% asphaltic content by weight. Small trenches and berms were constructed along the lot perimeter to ensure all runoff was captured. A 90° V-notch weir was installed below the lot to measure the discharge rates. Water quality samples were also obtained at the weir.

Lattice Block Lot

The interstices of the lattice blocks were filled with sandy loam and planted with bermuda grass. Runoff was monitored at a 90° V-notch weir located within a catch basin with an access manhole. Water quality samples were obtained at the weir.

Gravel Trench Lot

The study area consisted of a conventional asphalt lot with a 4-ft (1.2-m) wide drainage trench, ranging in depth from 2 to 4 ft (0.6 to 1.2 m) along the downslope width. The trench was lined with 6-mil polyethylene sheeting and filled with 1.5- to 2.5-in. (3.8- to 6.4-cm) diameter crushed stone which had been cleaned and washed. This base was topped with approximately 1 ft (30 cm) of smaller than 1-in (2.5-cm) diameter gravel. The trench was subsequently flushed with several truck loads of water to rinse out construction fines. Stormwater flows within the drainage trench were monitored within a 55-gal (208-L) drum. A 4-in. (10-cm) diameter pipe was used as the discharge control. A typical cross section is shown in Figure 3.

Conventional Asphalt-Concrete Lot

Runoff discharge was estimated from water levels at a 90° V-notch weir. Rainfall amounts during the natural storm events were obtained from a National Weather Service standard 8-in. (20-cm) weighing bucket recording gage located within 100 yds (91 m) from the site. Water quality samples were collected at the inlet to the weir.

Conventional Concrete Lot

Runoff volumes were estimated by staff gauge readings of the water level behind a 90° V-notch weir. Rainfall amounts were measured using a National Weather Service standard 8-in. (20-cm) weighing bucket recording gage situated on top of the City Hall Annex Building, located within 100 yds (91 m) from the lot. Water quality samples were obtained at the discharge from the weir.

9.2.4 Methodology

The lack of appreciable rainfall during the 26 months of sampling, and the rapid runoff response time of the lots were prime considerations for the decision to proceed with simulated rainfall testing. Sprinkler-induced "storms" provided the ability to control the intensity, duration, and timing of the rainfall events.

Sufficient antecedent dry periods were allowed between tests to allow accumulation of pollutants on the pavement surfaces. Impact-type sprinklers were used during the tests with the City's fire hydrants used as the source of water. The number of sprinkler heads was varied for each simulated storm and care was taken in the placement of the heads to provide uniform coverage of the lot. Equivalent rainfall estimates were obtained by placing eight wedge-type rain gauges on wooden stands around the test lot.

The gravel trench lot was too large for sprinkler coverage, so 2000-gal (7600-L) capacity rear-end-dispensing water trucks were used. The trucks drove slowly across the upper end of the lot releasing water at approximately 300 gal/minute (19 L/sec). Different event intensities were obtained by varying the number of trucks used, trips made, and number of trucks releasing at one time.

Sample collection, handling and analytical techniques employed in the study conform to recommended EPA[3] or American Public Health Association[4] methodology.

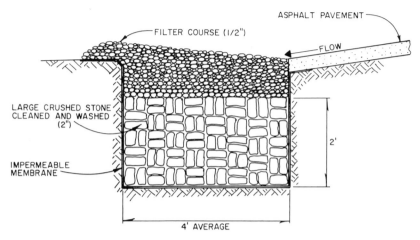

Figure 3.   Typical section of gravel trench.

9.2.5 Results of the Monitoring Surveys

The hydraulic and water quality results obtained during the stormwater surveys are discussed below with respect to hydraulic performance and intra-event pollutant characteristics. The inflow and discharge characteristics were calculated for each event. Subsurface runoff at the porous asphalt-concrete and gravel trench lots was monitored until flow ceased. These lots were open systems and did not prevent discharge, but rather, detained the discharge. Hence the runoff ratios cannot be used as an indicator of system storage performance.

The gravel trench was lined with an impermeable liner and the porous asphalt-concrete lot was constructed on relatively impervious limestone; hence, the only loss of inflow would occur in surface wetting of the pavement and base course material. The detention time was calculated as the difference in time between the center of mass of the inflow and the center of mass of the discharge hydrograph. A summary of hydrologic characteristics of the pavements during each runoff event is presented in Table 2.

Several studies have accumulated water quality information observed during urban stormwater events. Three classes of intra-event stormwater quality trends are frequently observed:

1. Initial high concentrations often occur at the beginning of the storm event, reflecting the initial removal of contaminants from the catchment. This phenomenon is termed "first-flush".
2. Variations in constituent concentrations may parallel the rising and descending flow rates of the storm hydrograph.
3. The absence of a well-defined intra-event trend in constituent concentrations may be observed.

A general water quality comparison between pervious and impervious lots, presenting flow-weighted average concentrations of total suspended solids (TSS), chemical oxygen demand (COD), total Kjeldahl nitrogen (TKN), total lead (Pb), and total zinc (Zn) is presented in Table 3.

Porous Asphalt Lot

As shown on Table 2, stormwater runoff conditions were generated at the porous asphalt lot during three sprinkler events. The highest average intensity of 1.67 in./h (4.2 cm/h) did not yield any surface runoff. The runoff ratio greater than unity for the second storm probably reflects a measurement error of the rainfall volume. As presented in Table 2, the total discharge volume, the time to peak flow, and the peak discharge rates were similar for each event even though the inflow varied from 0.50 in. (1.27 cm) to 1.53 in. (3.8 cm) and the average intensity varied from 0.48 in./h (1.27 cm/h) to 1.67 in./h (4.24 cm/h). For each event, the detention time was 42 min, indicating a uniform temporal response to the sprinkler inflows.

Levels of COD, TKN, and Zn in the underflow during the first and second events indicated a first-flush effect after the stormwater passed through the base layer. Values of TSS during the initial and final storms also reflected an initial rinsing of the pavement surface. The second event had the lowest sprinkler intensity during which time TSS and lead levels paralleled the discharge hydrograph.

Lattice Block Lot

Three simulated runoff events were monitored at the lattice block facility. The effective storage capacity of the lattice block lot, as estimated by the observed differences between rainfall and runoff, ranged from 0.64 to 0.82 in. (1.63 to 2.08 cm). The detention time of the lattice block lot was uniform for each event, ranging from 11 to 12 min. This rapid response time indicates that the runoff, which does not percolate into the pervious part of the pavement is transported off the lot within a duration comparable to impervious surfaces. The short detention times at the lattice block lot reflect nonuniform permeability distribution throughout the surface layer.

With regard to water quality, TSS and COD consistently displayed the first-flush effect while TKN levels paralleled the discharge hydrograph during two events. Metal concentrations did not display a discernable trend during the initial and final events, while a first-flush response was exhibited by zinc during the second event.

Gravel Trench Lot

Observed subsurface runoff from the three simulated storm events ranged from 64 to 77% of recorded volume, with an average of 73% for the three events.

TABLE 2. Summary of Hydrologic Data

Pavement Type	Event No	Total Inflow (in.)[a]	Duration (min)	Average Intensity (in./h)[c]	Peak Discharge (cfs)[b]	Time to Peak (min)	Total Discharge (in.)	Runoff/ Rainfall (in./in.)	Detention Time (min)	7-Day Antecedent Rainfall[e] (in.)
Porous asphalt	1	0.94	60	0.94	0.269	58	0.58	0.73	42	0.02
	2	0.50	62	0.48	0.253	54	0.64	1.28	42	0.09
	3	1.53	55	1.67	0.237	53	0.56	0.37	42	0.00
Lattice block	1	1.06	75	0.85	0.034	55	0.19	0.18	11	4.03
	2	1.08	60	1.08	0.078	40	0.39	0.36	12	0.00
	3	1.08	43	1.90	0.113	24	0.25	0.23	11	0.03
Gravel trench	1	0.64	94	0.41	0.440	60	0.49	0.76	29	4.03
	2	0.64	70	0.56	0.580	66	0.41	0.64	24	0.03
	3	0.64	59	0.65	1.667	55	0.49	0.77	19	0.12
Asphalt	1[d]	0.34	46	0.44	0.840	53	0.40	1.18	1	2.48
	2	0.21	10	1.26	0.223	7	0.15	0.71	5	0.99
Concrete	1[d]	0.85	120	0.43	0.200	58	0.46	0.55	18	0.53
	2[d]	0.57	33	1.04	0.100	30	0.28	0.48	14	2.48
	3[d]	0.45	90	0.30	0.070	30	0.17	0.38	17	3.71

[a] 1 in. = 2.54 cm.
[b] 1 cfs = 28.31 L/sec.
[c] 1 in./h = 2.54 cm/h.
[d] Denotes natural precipitation event. The remainder were sprinkler-induced events.
[e] Precipitation amounts recorded at the Austin airport within the indicated number of preceding days.

TABLE 3. Summary of Water Quality Testing Results

Pavement Type	No. of Sampling Events	Flow Weighted Average Concentration (mg/L)				
		TSS	COD	TKN	Lead	Zinc
Porus asphalt	3	175	25	2.2	0.014	0.031
Lattice block	3	25	33	1.6	0.012	0.020
Gravel trench	3	240	78	1.4	0.025	0.090
Conventional asphalt	2	43	30	1.3	0.012	0.012
Conventional concrete	3	14	20	0.5	0.021	0.007

Observations made during storm events indicated that the small-diameter surface gravel was impeding the vertical flux of water, i.e., a small portion of surface runoff was flowing across the top of the trench. Estimates of hydraulic detention times ranged from 19 to 29 min for the three events.

TSS, COD, Pb, TKN, and Zn levels paralleled the discharge hydrograph during the first and second events. A first-flush effect was observed for each constituent during the third event, which recorded the greatest average inflow intensity, 0.65 in./h (1.7 cm/h). The observed high suspended solids concentrations are possibly attributable to scouring of the backfill material supporting an asphalt ramp, which was constructed across the gravel trench before the initiation of stormwater sampling.

Conventional Asphalt-Concrete Lot

A natural precipitation event and a sprinkler-generated runoff event were monitored at this lot. A total of 0.34 in. (0.86 cm) of precipitation fell during the sampling period, with a maximum intensity of 2.32 in./h (5.9 cm/h). The sprinkler event was created by eight sprinklers operating for a duration of 10 min. A peak discharge of 0.22 cfs (6.23 L/sec) was observed with approximately 71% of the recorded inflow appearing as runoff. Low detention times were calculated for each event, indicating a rapid catchment response. The lack of depression storage on the lot, the small surface area and a relatively smooth surface contributed to the rapid response.

TSS and COD levels exhibited a first-flush effect during the two events. TKN values paralleled the discharge hydrograph during the first and second events. Both Pb and Zn concentrations paralleled the rising and falling limbs of the discharge hydrograph during the second event. The relatively high COD levels recorded at this lot, compared to the other impervious lot, may be attributed to the litter from adjacent deciduous trees which overhang the lot.

Conventional Concrete Lot

Three rainfall-generated stormwater surveys were conducted at this site. Observed runoff volumes ranged from 38 to 55% of recorded rainfall amounts, with an average of 46% for the three storm events. The runoff ratio was similar for each event and did not appear to be correlated to antecedent rainfall volumes. As with the pervious pavements, detention times for the conventional concrete lot were relatively uniform for each event, ranging from 14 to 18 min. These detention times are relatively long, compared to the other impervious lot, reflecting the storage of rainfall in surface abstractions prior to runoff commencement.

The water quality of the surface runoff leaving the concrete lot was better than was expected at the project outset. The lot is consistently filled to capacity with cars during weekdays and was not swept during the study period. During the initial storms, water quality values (except TSS and Pb) paralleled the discharge hydrograph; TSS and Pb did not show any discernible trends. TSS, TKN and Pb levels reflected a first-flush effect during the remaining events. A possible reason for the unusually low concentrations of contaminants is the amount of stormwater storage in surface depressions and subsequent sedimentation provided on the surface of the lot.

9.2.6 Comparison of Water Quality Results

A significant factor in determining water quality performance characteristics of the various pervious and impervious lots is the variable pollutant load present on the pavement at the beginning of each event. The contaminant load originating on each study pavement is a function of dry-weather accumulation rates and the stormwater hydraulic characteristics of each lot. Traffic type and density, presence of trees and other flora and fauna, as well as the interval of time since the last rain with intensities large enough to rinse the pavement, contribute to dry weather accumulation.

Total Suspended Solids

Concentrations of TSS demonstrated a first-flush effect during a majority of storm events at both pervious and impervious lots. Peak concentrations were generally greater in the runoff from pervious lots, possibly due to erosion of the diversion channels at the porous asphalt-concrete lot and flushing of the ramp construction fines at the gravel trench lot. Flow-weighted average TSS concentrations observed in runoff from the conventional asphalt-concrete lot were lower than that from the porous asphalt-concrete or gravel trench lot. Runoff from the conventional concrete lot was lower in TSS than from other lots.

Chemical Oxygen Demand

Intra-event trends of COD were markedly different between the pervious and impervious surfaces. Over 80% of the storm surveys on the pervious lots indicated a first-flush effect, while there was no predominant trend for COD on the impervious lots. As with the TSS values, peak concentrations for COD were generally greater on the pervious lots, possibly for the same reasons.

Total Nitrogen and Total Kjeldahl Nitrogen

Intra-event trends of TN were similar on pervious and impervious lots. There was no significant difference in average TN concentrations between the pervious and impervious surfaces. Average concentrations of TKN in the runoff from the conventional concrete lot were lower than in runoff from pervious facilities. Peak concentrations observed in runoff from the conventional asphalt-concrete lot were similar to peak values recorded for the pervious facilities.

Lead and Zinc

Intra-event trends of metals were similar for the pervious and impervious lots. Peak concentrations of zinc were significantly greater at the gravel trench lot while peak lead levels were anomalously high only at the conventional concrete lot during the initial storm event. The presence of metals in urban stormwater runoff is generally attributed to automobile pollution, including exhaust, fluid leaks, and mechanical wear. The gravel trench lot had the highest flow-weighted average concentration of both lead and zinc. This lot had the greatest automobile traffic per day; however, the high lead concentrations from the gravel trench site may not be directly attributed to the lot's traffic, which consists almost entirely of late-model cars using unleaded gas. Average zinc values were greater in the runoff from the pervious lots than from impervious lots.

9.3 RECOMMENDATIONS

Based on the results of this investigation, the following recommendations were developed:

1. The rapid response time of the study pavements and the stochastic nature of storm events made sampling of natural precipitation events quite difficult and labor cost intensive. As an alternative, future studies of this type should generate runoff events with sprinklers or other rainfall simulation devices.

2. Difficulties were encountered in obtaining uniform water distribution and accurate volumetric measurements when using the sprinkler network to simulate rainfall. For future simulated rainfall studies, use a greater sprinkler and rain gauge density, as well as an in-line flow meter to gauge the total inflow to the study pavement.

3. A Venturi-type flume, such as the Parshall flume, rather than a weir, should be used as a hydraulic control element for monitoring lot discharge. These flumes have the advantage of a lower head loss than a weir and smoother hydraulic flow, thereby preventing deposition of solids in front of the wier, such as was observed in this study.

4. Where municipal and watershed ordinances allow on-site depression storage on a pavement surface, use curbs in conjunction with porous asphalt pavement to increase hydrologic and water quality benefits. The presence of depression storage on impervious pavements appeared to provide a positive benefit for stormwater contaminant control. Periodic vacuum sweeping would remove accumulated contaminants and prevent their possible resuspension and discharge from subsequent stormwater scouring.

5. Information on the intensity, frequency and complexity of maintenance, as well as the useful life for various porous pavement facilities, should be compiled.

6. When employing gravel trenches a perimeter curb should be used to control off-site runoff. The permeability of the filter layer in the gravel trench in this study was exceeded by surface runoff, and a portion of the runoff volume bypassed the trench. An alternative recommendation is not to use the small-diameter filter course, which has a tendency to become clogged with silt.

7. Pollutant removal mechanisms in a porous pavement system have not been fully documented. The relatively slow hydrodynamics may allow some settling of suspended matter. Adsorption to, and absorption in, the base media may also be realized. Although transport of soluble constituents into the ground via infiltration removes them from the porous pavement facility, groundwater transport to receiving waters may result. Those impacts should be investigated.

REFERENCES

1. Goforth, G.F., E.V. Diniz and J.B. Rauhut. "Stormwater Hydrological Characteristics of Porous and Conventional Paving Systems," U.S. EPA Report-600/2-83-106, Municipal Environmental Research Laboratory, Office of Research and Development, Cincinnati, OH (1983).

2. Diniz, E.V. "Porous Pavement Phase I – Design and Operational Criteria," U.S. EPA Report-600/2-80-135, Municipal Environmental Research Laboratory, Office of Research and Development, Cincinnati, OH (1980).

3. "Methods of Chemical Analysis of Water Wastes," U.S. EPA Report-600/4-79-020 (1979).
4. "Standard Methods for the Examination of Water and Wastewater," 14th ed., American Public Health Association, Washington, D.C. (1975).

10   LIVING ON THE EDGE: ENVIRONMENTAL QUALITY IN THE COASTAL ZONE

I. INTRODUCTION

On a planet that is largely covered by water, the development of terrestrial species is confined to those habitats which can provide the basic ingredients for sustenance: an available food supply and potable water. During the millions of years since our ancestors crawled from the ocean, man has evolved into a creature that not only seeks to find a better habitat, but is capable of significantly altering that environment as he seeks to develop his community. Over the millennia, humans have spread across all of the available land masses, changing the landscape and exploiting the terrestrial resources to meet the demands of an ever-evolving and demanding society.

After some four and a half million years of evolution, we find ourselves distributed rather unevenly across the land surface, with the vast bulk of our population concentrated on the edge of land. Within the U.S., some 110 million people, about one half of the national population, now live within coastal drainage areas (Figure 1). Within the next two decades, this coastal population in the U.S. is projected to increase by an additional 47 million people. Thus the coastal zone is under dramatically increasing pressure from population growth and relocation, including the alterations of land use and vegetative cover which have accompanied this migration pattern.

One of the most important environmental factors that has attracted our species to the edge of land is the tremendous biological productivity in this "coastal zone" between land and water. In the relatively narrow aquatic habitat where mineral nutrients and organic matter are washed from the land into the ocean, the richest and most diverse ecosystem on the planet has developed. Along the eastern coastline of North America lies a series of estuaries, barrier islands, and

0-87371-805-4/93/$0.00+$.50
© 1993 by Lewis Publishers, Inc.

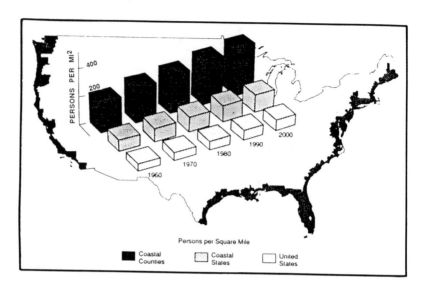

Figure 1. **Coastal counties in the United States – population and density growth.[1]**

a shallow marine shelf which have created one of the most productive finfish and shellfish habitats in the world (Figure 2). This resource has been extensively exploited by human occupation and development, and we now find that ecosystem under threat from a combination of factors, especially pollutant inputs. The coastal zone from the New York City (NYC) metropolitan area south along the coast of New Jersey (Figure 3) has borne witness to much of this coastal pollution, including oil spills, coastal dumping, and inadequately treated wastewaters. The greatest threat to environmental quality in the New Jersey Coastal Zone, however, is continuing land development within the 5400 km^2 (2086 mi^2) of mainland and barrier island drainage to these coastal waters, which are described as the New Jersey Atlantic Coastal Drainage area (ACD).

10.2 WATER QUALITY

In the New Jersey ACD area, land-use alteration and increased population have caused significant degradation of water quality from an increase in both point source and nonpoint source (NPS) pollutants. While point sources of pollution are primarily the sewage treatment plants, NPS pollutants scoured from the land surface and flushed into coastal waters with each rainfall comprise a largely unmeasured and unmanaged flux of contaminants and contribute significantly to the declining water quality. The focus of this paper is to quantify the sources and types of NPS pollution in the New Jersey ACD, and to describe the specific guidelines and measures being undertaken by the State of New Jersey for the control of this NPS pollution, from both existing development and new growth.

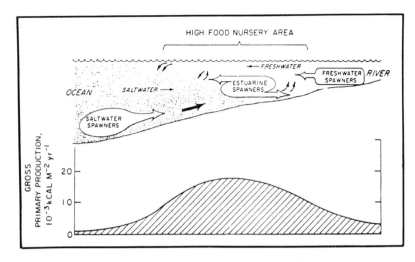

Figure 2.   Primary productivity in coastal estuaries.[2]

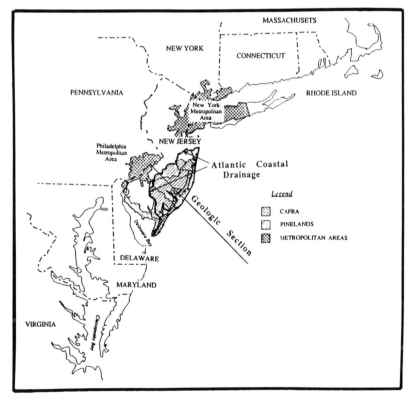

Figure 3.   Atlantic coastal drainage (ACD) along the New Jersey coast.

It is recognized that managing the sources, quantity, and effects of point source pollution in the coastal zone is of paramount importance in improving coastal water quality, especially with respect to the nutrients, nitrogen and phosphorus, which pass through our wastewater treatment plants largely untreated. In addition to the New Jersey P.S. discharges, NYC alone discharges over 2 billion gal/d of wastewater effluent, 10% of which is raw sewage. During rainstorms, a combined sanitary and storm sewer system overwhelms the treatment plants and some 650 combined sewage outfalls that ring New York harbor discharge twice the daily amount as untreated wastewater. This discharge of the Hudson-Raritan estuary flows from the harbor and south along the coast, mixing with ocean waters. Sewage treatment plants in the New Jersey Coastal Zone discharge another 128 million gal/d of treated wastewaters to these ocean waters, so it must be recognized that improving water quality in the coastal waters depends on addressing point source discharges.

The perspective of this paper, however, is the impact of pollution from stormwater discharges, and how these discharges contribute to the loss of environmental quality. Specifically, how the patterns of urban development within the coastal zone, and the regulations which attempt to control it, affect NPS pollution.

In this relatively fragile ecosystem, managing stormwater is an essential part of any plan for continued use of the coastal habitat. The population growth that has produced these pollutant increases is largely attributable to the high quality of life found in the coastal zone, which in turn is vulnerable to the water quality degradation caused by the organic matter, nutrients, heavy metals, and synthetic organic chemicals which we generate by our land use activities (Table 1). Continued use of this coastal ecosystem as a resource requires an immediate and long-term solution to controlling these pollutant impacts even as the population and the demands on the coastal zone increase.

10.3 NEW JERSEY COASTAL AREA: ENVIRONMENTAL SETTING

The New Jersey ACD consists of approximately 5402 km^2 (2086 mi^2) of land and water, including some 129 km^2 (50 mi^2) of barrier islands, 738 km^2 (285 mi^2) of wetlands, bays and estuaries, and some 4532 km^2 (1750 mi^2) of mainland, covered largely in pitch pine-cedar forest. This flat coastal plain (see Figure 3) is comprised of a series of unconsolidated sedimentary deposits of sand, marl, and clay, which increase in thickness toward the coastline (Figure 4). Over the past 16,000 years, as the ocean level has risen, the water's edge has progressed inland to it's present position. The ocean currents and upland erosion and deposition have created a long, narrow series of barrier islands, which absorb the energies of ocean storms and buffer the estuary habitats from the scour of waves and currents. Between the mainland and barrier islands, there exist embayments and estuaries of different sizes and configurations where, in many areas, inland

TABLE 1. NPS Pollutants in the New Jersey Coastal
Zone Stormwater Management Criteria (mg/L)
(Cahill and Associates, 1988)

Pollutant	Stormwater Criteria	Present Levels
Total phosphorus	0.050	0.030–1.00
Total nitrogen	0.700	0.300–2.00
COD	10.000	5–150
Total suspended solids	20.000	10–500
Enterococci	35 CFU/100 mL	Unknown
Petroleum hydrocarbons	1.000	2–15
Metals[a]		
Lead	0.020	0.010–0.700
Mercury	0.001	Unknown
Arsenic	0.010	Unknown
Cadmium	0.010	Unknown
Synthetic organics		
MAHs	0.100[b]	Unknown
PAHs	0.100	Unknown

[a] See discussion of heavy metals criteria in text.
[b] Several of the halogenated and monocyclic aromatic hydrocarbons included in this
grouping have specific standards in drinking water, including trichloroethylene
(0.005), carbon tetrachloride (0.005), vinyl chloride (0.002), dichloroethane (0.005),
and benzene (0.005).

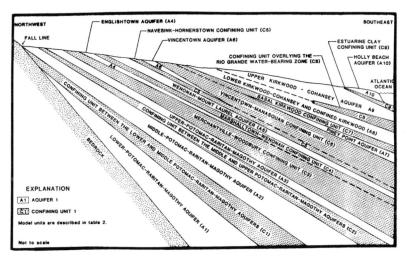

Figure 4.   Generalized hydrogeologic section of the New Jersey coastal
plain.[3]

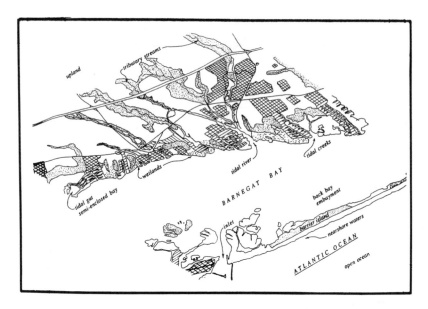

Figure 5.   Coastal zone habitats.[4]

erosion and marine sediments have gradually filled to create extensive wetlands (Figure 5).

The interior of the flat, sandy mainland is extensively covered by a scrubby pitch pine and cedar forest, known locally as the "Pine Barrens", creating yet another unique aspect of this coastal zone. In this coastal drainage area, the hydrologic cycle is quite different from inland watersheds. Of the 114 cm (45 in.) of annual rainfall, only a small fraction occurs as direct runoff (6.35 cm/year or 2.5 in./year), while the balance infiltrates rapidly and moves laterally to shallow surface waters. The unconsolidated sand strata allow rapid infiltration (50 cm/year or 20 in./year) to the shallow groundwater table, most of which (43 cm/year or 17 in./year) discharges to surface streams. The result is a longer, more attenuated hydrograph, with subsurface flow playing a far greater role than in upland watersheds. As this area is altered by development,increased impervious areas and changing flow pathways (inlets and storm sewers) convey pollutants introduced by development directly to the coastal waters. In addition, fresh water recharge to the underlying aquifer is reduced, and the resulting increase in saltwater intrusion into the sand aquifers contaminates water supply wells along the coast.

10.4 POLLUTION OF COASTAL WATERS: SOURCES AND QUANTITIES

In the coastal waters of New Jersey, the primary impact of urban stormwater runoff is not an increase in flooding problems, but rather a decrease in water quality, which directly effects the coastal waters. In addition, the loss of the stormwaters as a source of groundwater recharge is compounded by the increased groundwater withdrawals for new water supply. The past decade has also seen a massive effort to collect and treat all sewage produced by coastal development, with long outfalls discharging the effluents to nearshore ocean waters beyond the barrier islands. This has exported the groundwater withdrawals beyond local recharge and has transferred the pollutant impact from the estuaries to the nearshore ocean waters, where the nutrients enrich the waters and support massive algal blooms. Increased development along the mainland area continues to pave over the sandy soil and add NPS pollutants to estuary drainage via stormwater, offsetting the local benefit of sewage conveyance to the ocean. The end result is a pattern of urbanized land along the barrier islands and mainland (Figure 6), with 1.13 million permanent residents (and still growing), and an additional 1.5 million summer tourists. This urban growth has dramatically altered the natural drainage system (and the landscape) in a way that is significantly increasing the discharge of pollutants from the land surface to coastal waters.

Before steps can be taken to improve water quality in the coastal zone, the specificconditions causing the water quality degradation must be better understood. In the New Jersey coastal waters, much emphasis has been placed on coliform bacteria levels as a measure of water quality, and the measurement of other pollutants, especially nutrients, has been quite limited. With the regionalization and upgrading in wastewater treatment facilities to secondary treatment with ocean discharge, the bacteria levels in the bays and estuaries have been reduced. However, this does not necessarily indicate improved water quality for the ecosystem. It is also known that the coastal waters have experienced excessive algal growth in recent years, with the key pollutants in question being the nutrients nitrogen and phosphorus. One needs only to consider the manicured lawns and golf courses which cover the once sparsely-vegetated landscape to realize that a significant part of our chemical alteration of the environment is reflected not on the land, but in the water.

The State of New Jersey has implemented the development of a computerized geographic information system (GIS) for environmental analysis and resource planning, including quantification of NPS pollution. Current research by Cahill Associates and the NJDEP& E[5] has developed a data file of urban land use within the coastal drainage, based on land use, impervious cover and NPS pollutant production (estimated in part from fertilizer application data). Figure 6 shows the application of a computerized database used in the analysis of land use in the coastal zone. A key issue here is that NPS pollution is not only a function of impervious cover, but that the intensity of lawn care and maintenance is also significant. An urban landscape with highly maintained lawn areas can be a significant source of nitrogen and phosphorus in stormwaters. Some 2500 polygons of different types of urban land use (residential, commercial, recreational, etc.), representing about 10% of the New Jersey ACD area, are currently

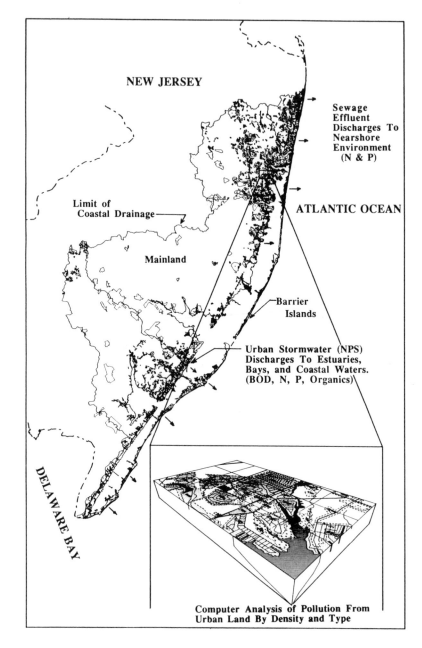

Figure 6.   Urban land in the coastal zone.[5]

being studied. Based on this information and prior estimates of land drainage inputs, a comparison of the relative contribution from both point and nonpoint sources of pollution to coastal waters is presented in Table 2. Although urban land use accounts for only 15% of the land use in the ACD, 26% of the NPS pollutant loadings of nitrogen and phosphorus to coastal waters are derived from urban lands. More importantly, over 90% of the nutrient loadings to the bays and estuaries, currently experiencing enrichment problems, are due to NPS loadings. This very conservative estimate assumes that only a portion (10%) of the nitrogen and phosphorus (4%) applied as fertilizers actually find their way into coastal waters. Due to the proximity of these urbanized lands to the waters edge, the likelihood of these pollutants entering into the aquatic ecosystem is much greater, but scientific evidence of this theory is lacking. Future chemical sampling may indicate that these urban NPS pollutant inputs situated along the fringe of coastline are much greater than estimated here. With the projected future growth in the ACD, even a conservative analysis indicates the importance of controlling NPS pollution from urban lands. It also illustrates the power of the GIS for environmental studies of existing conditions, and lays the groundwork for analysis of future growth impacts. Such a database facilitates the development of plans and programs to minimize further water quality degradation. It also has allowed the selection of appropriate techniques to reduce or control NPS pollution, from both existing and future urban lands (Figure 7).

10.5 PUBLIC PERCEPTION AND LAND-USE REGULATION

In New Jersey, attempts to reduce NPS pollution have consisted primarily of land-use regulations and, more recently, promoting the implementation of "Best Management Practices" (BMPs) for stormwater management which reduce the pollutant load in these stormwaters. As the coastal region has grown and changed over the past three decades, the water quality impacts have led to a public outcry for protection of the environment, and the passage of several regulatory controls on land use. The public perception of the problem is based on a series of seemingly unrelated incidents which create a sense of loss, and which significantly affect the recreational and other aspects of the coastal area. In the past few years, the "Jersey Shore" has experienced a number of serious environmental problems which received widespread media attention across the U.S. These problems include a declining water quality emphasized by beach closings, unexplained dolphin deaths, floating garbage, intense phytoplankton growths ("red", "green", and "brown tides"), and beaches strewn with "tar balls" and medical wastes. In addition, the finfish and particularly the shellfish harvests have significantly declined or been restricted, and the filling of wetlands and clearing of woodland have destroyed major portions of wildlife habitat along the coast.

TABLE 2. Pollution Sources to the New Jersey Coastal Zone

Inland Drainage to Bays and Estuaries (4,532 km²)

Nonpoint	Area (km²)	% of Total	Nitrogen (mton/yr)	% of Total	Phosphorus (mton/yr)	% of Total
Agricultural	432	9.53	912	44.04	87	48.60
Forest	3107	68.56	465	22.45	31	17.32
Urban, coastal	500	11.03	464	22.40	45	25.14
Urban, parkland[a]	9	0.20	5	0.24	1	0.56
Urban, Pinelands[b]	153	3.38	61	2.95	2	1.12
Misc.[a]	29	7.26	164	7.92	13	7.26
Marinas	2	0.04				
Total Nonpoint	4532		2071		179	
Point Sources to Inland Waters			82		19	
Total to Inland Drainage			2153	96.16	198	90.40

Wastewater Effluent Discharges to Ocean (From ACD)

	Nitrogen		Phosphorus	
Point Sources to Nearshore Waters (14 major STPs along coastline)	5196		1,203	
Total:	7349		1,401	
(NPS)		28.18		12.77

Total urban drainage area is 14.61% of the ACD, contributing 25.59% of the NPS nitrogen load, and 26.82% of the phosphorus load.

[a] Based on 5 kg/ha/yr TN and 0.4 kg/ha/yr TP.
[b] Based on 4 kg/ha/yr TN and 0.1 kg/ha/yr TP.

In 1988, a *Blue Ribbon Panel on Ocean Incidents*, aided by the New Jersey Department of Environmental Protection,[6] found that ..."Development was singled out as a major cause of pollution to the marine environment." The panel recommended ..." eliminating federally guaranteed flood and storm insurance in sensitive coastal areas and limiting growth in the coastal zone." To this end, the State of New Jersey has been attempting to control development, and it is essential that this new growth be guided and managed in such a manner as to reduce pollutant discharges to coastal waters. Along the coast, a set of regula-

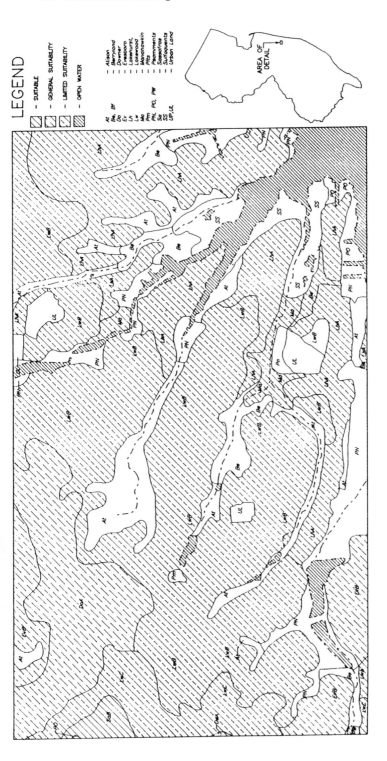

Figure 7.   Evaluation of BMP suitability by use of GIS model.[5]

tions was adopted some 18 years ago which reviews new development and limits the type of development and amount of impervious cover in certain areas. This Coastal Area Facilities Review Act (CAFRA) has had some effect, but numerous loopholes have been found to circumvent the intent of the law. In the inland drainage, the sensitive Pinelands have been protected by special legislation of a decade ago. The end result of that law is to direct and channel all new growth into limited high growth regions closer to the coast, which will produce their own set of coastal water quality impacts. At the same time, agricultural land use on the mainland is given preferential treatment and protection, even though it contributes significantly to the pollutant loading of coastal waters. While less than a perfect system, this regulation has managed to prevent any new incursions of development into the heart of the Pinelands ecosystem.

Thus, one of the strategies of the past decade has been to attempt a reduction of pollutant inputs by partially controlling the location and density of development, and influencing the intensity of land use in the coastal zone. This program has had some effectiveness, especially in terms of preventing further filling of wetlands and other sensitive habitats. However, it has not reduced the pollutant inputs from development, and it has become obvious that land-use controls alone will not reverse deteriorating water quality. A coastal zone water quality management program must include the management of stormwater quality. This does not imply that massive new treatment works are appropriate at the end of our storm sewers, but rather that we use specific measures and construction techniques, on a site-specific basis, to reduce the quantity of pollutants contained in the stormwaters which are generated. That is, the regulatory framework must contain both "how to build" guidelines, as well as "where not to build" guidelines. The effectiveness of such NPS pollution reduction measures is very much a function of natural conditions at any given point in the coastal environment, such as soils and depth to water table, but the basic approach is one of defining the tolerance limits of the environment, and learning to live within them.

Much of this policy was contained in a report of April 1989 by Cahill Associates, [4] titled *Stormwater Management in the New Jersey Coastal Zone*, which proposed a number of methods for reduction of NPS pollutants from stormwater, as well as programs for reduction of disturbance in the development process. These structural BMPs include measures which would prevent excessive site disturbance (described as a policy of minimum disturbance and minimum maintenance), use of special materials for reduction of stormwater runoff (porous pavement and groundwater recharge), stormwater treatment systems (water quality detention basins, artificial wetlands) and other development guidelines. The application of these guidelines for further new development appears to be gaining public support and should be translated into regulation in the near future, but the problem of how to retrofit (or redesign) the existing land development (and associated stormwater infrastructure) is a problem yet to be solved.

10.6 CONCLUSIONS

The analysis and applied research discussed here indicates that the state should rethink the current policy of wastewater collection, secondary treatment, and discharge to nearshore coastal waters, since the resultant nutrient input is enriching these waters and contributing to the algi bloom problem. In addition, the state must regulate (and significantly reduce) fertilization practices in the coastal zone, especially where these artificial landscapes are in close proximity to the water's edge. Measures must be required for NPS reduction where new development goes forward, and the entire process of land development must be rethought. Finally, the land-use policies reflected in certain programs, such as the Pinelands region, must temper the impact of proposed high-density zones, especially where they drain directly to coastal waters. The limits of the freshwater resource, in terms of both quantity and quality, will be exceeded by the current management program.

As more of our population migrates to the coastal zone, the importance of protecting that fragile ecosystem takes on new emphasis. The pollutants we generate by our efforts to alter that habitat, as well as the wastes we produce, must be prevented from entering these waters and destroying the natural balance that exists between land and water. The concept of stormwater management takes on an entirely different meaning when viewed as one of the mechanisms of this pollutant transport. For centuries we have built engineering works along the shoreline to protect our structures from the ravages of ocean storms, but now we must create systems to protect the ocean water from the ravages of our structures. Understanding where we live should guide how we live, and as we crowd along the land's edge in increasing numbers, we must take special care to protect the waters that we share.

REFERENCES

1. "Selected characteristics in coastal states," 1980–2000, U. S. Dept. of Commerce, National Oceanic and Atmospheric Administration, Strategic Assessment Branch, Ocean Assessments Division, Rockville, MD (October, 1989).
2. Clark, J. "Coastal ecosystems: ecological considerations for management of the Coastal Zone," The Conservation Foundation, Washington, D.C. (1977).
3. Martin, M. "Ground-water flow in the New Jersey coastal plain," U. S. Geological Survey Open file Report 87-528, West Trenton, NJ (1989).
4. Cahill Associates (CA). "Stormwater management in the New Jersey Coastal Zone," New Jersey Department of Environmental Protection and Energy, Trenton, NJ (1989).

5. Cahill Associates. "Limiting NPS pollution from new development in the New Jersey Coastal Zone," New Jersey Department of Environmental Protection and Energy, Trenton, NJ (1991).
6. "The state of the ocean," a report by the Blue Ribbon Panel in ocean incidents, New Jersey Department of Environmental Protection and Energy, Trenton, NJ (1988).

11   A POLLUTION CONTROL MANAGEMENT PLAN FOR THE TOWNS OF DOUGLAS AND ONCHAN, ISLE OF MAN

11.1 Introduction

The Isle of Man is an independent state located in the Irish Sea, with an overall population of 75,000 (see Figures 1 and 2). It is part of the United Kingdom but at present is not a member of the European Community (EC). The Island Government has a policy of adherence to the laws of its neighbors which has resulted in the need for a pollution control management plan to be formulated. The objective is to meet both the highest standards of public health and environmental protection for the island and also to comply with all relevant EC directives or other legislation, particularly bathing beach quality and fisheries legislation.

The Borough of Douglas and Onchan District Commissions jointly sought proposals to advise upon the sewerage disposal arrangements for the main communities of Douglas and Onchan (peak population, 55,000), precipitated by real concern for the pollution to sensitive areas of the foreshore, the bathing beach, and fishing industry. This, combined with the uncertain physical condition and operational deficiencies of the existing infrastructure together with the need to attain discharge standards in line with surrounding countries, has highlighted a number of distinct pollution problems. Existing drainage systems are becoming stressed and modifications to the system are required. The standard EC classification for amenity beach areas is now becoming widely known, and given the importance of tourism within the Manx economy and the Isle of Man's reputation for providing an attractive environment, failure to achieve targets

0-87371-805-4/93/$0.00+$.50
© 1993 by Lewis Publishers, Inc.

Figure 1.   The Isle of Man's location in the Irish Sea.

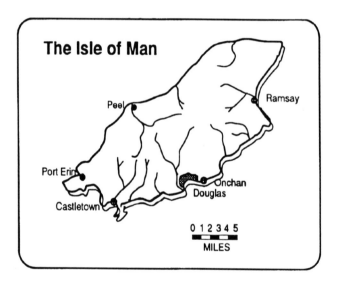

Figure 2.   The Isle of Man.

compatible with this objective, could have detrimental effects on the island community when compared with developments in neighboring countries.

Douglas is a densely-populated seaside resort with harbor, beach, shopping, and recreational facilities. Tourist accommodation is concentrated close to, and parallel to the promenade and in the south of the town, whereas the residential

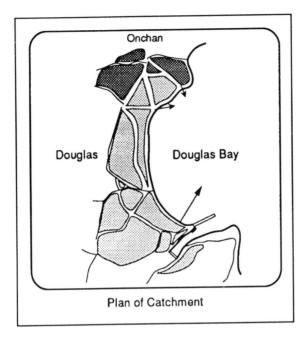

Figure 3.   Plan of catchment.

accommodation is distributed more evenly over the urban area of Douglas and the residential side of Onchan. The high-density developments are in the older part of Douglas and in the area to the north of the harbor. It is in this area that the original sewerage system for Douglas evolved.

At present the developed area (bounded to the east by the sea, the south and the west by the rivers Douglas and Glass, and to the north by the river Groundle) occupies an area of land that rises toward the northeast from Douglas towards Onchan, being cut by several streams running east or west which have cut steep valleys that are used as parks and that, during storms, act as overland floodways. Douglas Bay, an attractive sandy bay protected from the worst of the weather by headlands north and south, is now fronted by a promenade and is very popular with visitors. The town is continuing to grow and is now beginning to overspill the natural boundaries (Figure 3).

The drainage of Douglas and Onchan developed from a Victorian sewerage system. The main Douglas system has operated satisfactorily for some 80 years and is concentrated around a high- and low-level foul drainage network and separate storm drainage system. This discharges through a cast iron outfall into Douglas Bay. Separate catchments with formalized sewerage systems and associated outfalls have been constructed for the communities of North Douglas and Onchan (Figure 4).

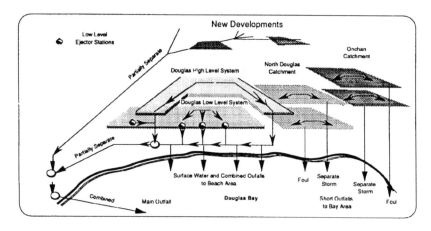

Figure 4.   Douglas and Onchan existing sewerage.

Recent developments in the sewerage network have resulted in some loss of the completely separate system with increased quantities of surface water being directed to the foul system in areas where the surface water system has not been extended. Additionally seawater inundation has affected the ability of the network to discharge freely at all times, and the overall result is that pollution of the foreshore has increased. The absence of detailed sewer records has further served to emphasise the need for investigations and arising from this a number of studies have been undertaken over the past four years by Acer Consultants to investigate both the level and the source of the pollution to the foreshore area and associated bathing waters. These previous studies have identified deficiencies in the drainage system, the need for treatment, and the relocation of the discharge point of the wastewaters.

The basis behind any rationalization of the sewerage system is that discharge of wastewaters into inland watercourses could not be provided within the boundaries of Douglas and Onchan, and it is difficult to justify the provision of the sophisticated treatment processes involved or to take the responsibility for the care of the river. Therefore collection of the community wastewaters was proposed to be directed to a central location and then to be pumped inland for treatment. Final discharge of treated wastes would be directly through an extended outfall for release into the well oxygenated waters outside Douglas Bay. The final treatment strategy for the wastewaters is being reassessed at the time of writing, and investigations are underway to determine whether collection of all the island's wastewaters and delivery through high pressure pumping mains, incorporating chemical treatment, to a central collection and treatment plant would be economically attractive and fulfil the environmental standards for the Island.

Historically the communities of Douglas and Onchan have always discharged wastewaters into Douglas Bay and the surrounding coves through marine outfalls with no treatment applied to the discharge. The community is divided into three well defined drainage catchments with their associated outfalls for foul flows located at Conister Head and Port Jack Bay, with a number of smaller outlets on the Onchan headland. Runoff resulting from surface waters are conducted overland or through the surface systems to discharge through short outfalls or directly from the promenade onto the beach area.

The main commercial and tourist area of Douglas lies in the flat area behind the promenade extending from the port area in the south to the rising ground to Onchan. The high ground area behind the main Douglas Community and in the Onchan area is divided by several small rivers, principally the River Groudle to the north and the rivers Dhoo and Glass to the south.

Several features of the town from a drainage point of view must be noted:

- It is a feature of the town, for which the original developers should be congratulated, that there are remarkably few road gulleys in the older roads of the town. As the highways generally slope (steeply at times) to the sea or to a river they have clearly served the area as a major overland storm drainage network with surface storm flows causing little perceived problems to the locals.

- The main pedestrian precinct shopping area in the town and the quay area are now at a lower level than any overland flow outlet, due to the construction of the Promenade. The combined sewers serving this area will need to continue to provide free discharge or this area will, by default, become an aboveground stormwater storage area and flood.

- Inevitably there would seem to be some illegal cross connections in the areas served by separate systems, with high levels of fecal coliforms (1×10^6/100 ml) being measured in the stormwater. In addition the stormwater itself is polluted during its travel over the land, which will wash off dog, bird, or animal waste, grass cuttings, etc., so it would seem that pollution directly attributable to the stormwater sewers is affecting the bathing beaches and the benthos of the local rivers and streams.

- A long standing problem with the area has been the absence of an accurate set of sewerage records to assist development strategies. Recent construction has therefore proceeded on a piecemeal basis to meet new developments with remedial works to either overcome the surcharging effects within the system or to restrict the infiltration from ground- or seawater. The establishment of firm records is therefore considered of vital importance and, together with the remedial works program, will reduce the load on the sewerage system and minimize the volume of sewage to be transported, specifically as shock loads during periods of high rainfall. The development of records is proceeding and will not conflict with the strategies presented.

- It is also suspected that the system is affected by seawater intrusion, particularly during periods of high tide.

In general the problems of the area fall into the following categories.

11.1.1 QUANTITY CONSIDERATIONS

In general, stormwater flooding is not a problem. The system seems to cope well with intense rainfalls with only isolated incidence of surface flooding that rarely affects property. Some basement flooding does occur but much of this is suspected to be due to direct seepage from the sea. Generally, it has been concluded that the present level of service provided by the system is acceptable to the local residents, and the management plan is aimed to maintain present standards as the town develops in the future, especially as any degredation in the level of service could affect the tourist trade on the island.

In common with many towns as they have grown, the drainage system developed piecemeal in order to handle each extension resulting in several distinctly different systems being in use. The older part of Douglas is located on a low-level area near the harbor drained by simple egg-shaped combined sewers that collect wastewater to discharge to a short sea outfall into the bay. Basement drainage in the lower part of the town was added to the 19th century developments as a low-level system that is pumped by using sewage ejector pumps to the short sea outfall.

11.1.2 Quality Considerations

- Bathing beach quality: the existing outfall discharge within Douglas Bay and the tidal pattern are such that the untreated discharge is brought right onto the beach at certain states of the tide. A visible sewage slick often forms, so even a layman can see the sewage washing towards the beach.
- Storm water overflows: These are located at the points of overload on the system and operate regularly, some virtually whenever it rains, and contribute considerably to the pollution of the bathing beach.
- Surface water discharges: These are causing pollution to the beach, and causing some damage to the local streams and rivers.

The management plan is therefore aimed at a phased scheme to tackle these problems by redirecting the foul sewage to a sewage treatment plant, and to use source control techniques and attenuation storage to reduce and eliminate the operation of the storm water overflows while reducing the escape of pollution from the present separate stormwater sewers (Figure 5).

11.1.3 Objectives

In accordance with the Island Government's policy of adherence to the standards of their neighboring countries, the relevant EC directives provide

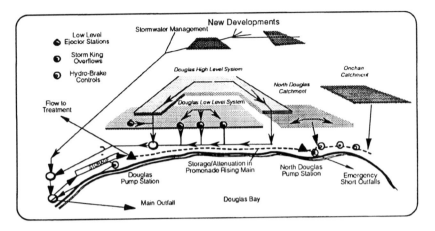

Figure 5.   Douglas and Onchan proposed system.

guidelines for the necessary performance in the future. Without reproducing these in full, there are three main targets that need to be met.

1. the EC standards for bathing beaches be complied with, particularly the number of fecal bacteria that may be present where people are bathing
2. the proposals of the draft EC directive on municipal wastewater treatment be complied with, which requires treatment on all sewage discharges before a sea outfall to a minimum of primary treatment of the nonsensitive wastes, and secondary treatment if the waters are sensitive
3. the general principal that pollution be controlled so as to meet use-related standards in any receiving waters

It is the intention of the scheme to reduce the flow presently carried by the existing system. Rehabilitation work currently underway has the dual effect of both reducing the groundwater infiltration and serving to increase the storage capacity within the existing system.

A measure of the effectiveness of any new system will be the capability of achieving the development targets through well-defined stages such that previous capital investments are protected and the performance of the existing installation is not seriously disrupted.

The existing system does contain storm overflow arrangements discharging from the foul to the surface water system to ensure its continued operation during periods of extreme storm conditions. It is intended that these overflows into the surface water system will eventually be eliminated by the provision of additional upstream storage within the catchment created from the attenuation of foul flows within the system by the incorporation of Hydro-brake™ flow controls.

11.1.4 Investigations Carried Out

In view of the absence of data on the drainage system, the following studies were necessary to enable the areas of concern to be identified, the issues to be clarified, and a coherent plan to be prepared:

- tidal current and sea outfall studies to determine the tidal currents and investigate the feasibility of sea disposal of sewage
- sewage system investigations using flow monitors to determine the foul flows in each system, degree of stormwater infiltration, etc.
- the building of a mathematical model of the main sewers in the system to determine the capacity of the system and locations of hydraulic shortfall using a WASSP model.
- closed-circuit T.V. surveys of the main services to determine the system's structural integrity, and enable a rehabilitation program (particularly on the brickwork egg-shaped sewers) to commence
- hydrologic analyses to enable the shortcomings of the system to be identified, and to be used to identify optimum strategies

11.1.5 Alternative Strategies

Using the available data several alternative strategies for the area could then be considered, including as follows:

- Conventional uprating of the separate sewage systems, with multiple long sea outfalls — a strategy clearly both expensive and incapable of meeting the objectives.
- Conventional uprating of the separate sewage systems, with interception of the outlying catchments to a single long sea outfall — also costly and unlikely to meet the objectives.
- Interception and discharge to a land treatment plant, with return of treated effluent to a short sea outfall — still costly and probably ineffective.
- Use of attenuation storage to reduce pump flows and the incidence of overflow, with interception of the outfalls to discharge to an inland sewage works, and the effluent being discharged back to a long sea outfall — would be the recommended scheme for the area of Douglas and Onchan in isolation.
- Include the area in a strategic pollution control system for the complete island. This optimum scheme uses attenuation to reduce pump flows and stormwater discharges, and intercepts all the foul sewage to be discharged to a new inland sewage treatment plant serving the complete island.

11.2 GENERAL CATCHMENT STATISTICS

11.2.1 Population

Catchment	Residents	Visitors	Summer Peak
Onchan	9,150	1,200	10,350
North Douglas	5,950	5,400	11,350
Douglas	18,400	17,100	35,500
Totals	33,500	23,700	57,200

11.3 DESIGN RATES OF FOUL FLOW

The existing drainage system was considered to be a combined system with the overall contribution to the flow determined from Formula "A" as produced from the *Technical Committee on Disposal of Storm Sewage*. The contributions to each of the catchments in relation to the population served is as follows:

Catchment	Population	Demand (L/d)	Foul Flow	Infiltration (L/sec)	DWF (L/sec)	Storm-water Contribution	Formula Flow "A" (L/sec)
Onchan	10,350	200	24	3	27	162	189
North Douglas	11,350	200	26	3	29	177	207
Douglas	35,500	200	82	9	91	555	696
Totals	57,200		132	15	147	894	1092

11.4 DESIGN RATES OF FLOW FOR TREATMENT

Treatment of the flows will be controlled from the rate to be pumped which must be a balance between the foul contribution and storm and infiltration totals.

Catchment	Formula "A" (L/sec)	3 x DWF (L/sec)
Onchan	189	81
North Douglas	206	87
Douglas	696	273
Totals	1091	441

11.5 FLOW ATTENUATION

Flow attenuation using storage and source control involves modification of a drainage system so that overloading of any downstream areas is minimised and the action of any emergency overflows from the system is restricted to the minimum. For any existing system this can be achieved by mobilizing available storage in the large-diameter pipes and by regulating the flows passing downstream from smaller diameter pipework at the head of the system. For new systems, a design of the installation by means of network analysis should be completed allowing for all reasonable peak loads.

System control is undertaken by a combination of vortex flow controls that restrict flows passing downstream through the network to defined limits, and the incorporation of storage facilities, both underground and/or in surface storage areas where acceptable. Control of flow can be effected by regulating the output passing downstream using flow controls to ensure unimpeded dry-weather flow can enter from lower areas of the network while preventing overloading of the sewers in storm conditions.

Vortex flow controls have been developed for installation in access holes or similar structures in new or existing systems, which, with no moving parts, will provide accurate and reliable control of flow. At low heads, the vortex will not form and free-flow conditions in the system prevail. Proprietary devices such as the Hydro-Brake Flow Control™ (Reg-u-Flo® Vortex valve in the U.S.), incorporate no moving parts, consume no energy, require no sophisticated sensors, and reduce maintenance commitments.

Any installation must be designed such that the operation is reliable when called into operation in the most severe conditions. Their installation requires the minimum of maintenance and the operation is easily understood by all members of the operational staff.

11.6 STORMWATER OVERFLOWS

Storm King™ Overflows take advantage of the naturally occurring phenomenon of *dynamic separation* to divert pollutants to the foul sewer while comparatively clean water overflows to the outfall. These devices have been demonstrated to be highly efficient at reducing the pollution arising from stormwater overflows. Their use will enable the existing overflows to be retained during the early phases of the scheme while ensuring that the objectives of pollution control are met, and as the source control system is developed their frequency of operation will reduce.

11.7 SUMMARY OF STRATEGY

Using these techiques, the recommended strategy for the area falls into several clearly defined categories:

1. At present the storm time behavior of the system is held to be satisfactory with the exception of pollution escape in times of storms.

2. Within a phased scheme, in phase 1, it is assumed that all existing storm overflows will be retained, but uprated where possible.

3. At the proposed pumping stations sites, it is assumed that a frequency of operation of the overflows as dictated by Formula "A" will be acceptable.

4. The reduction of pumped flows to treatment to 3 x DWF. The difference between 3 x DWF and Formula '"A" overflow setting is to be balanced by the provision of attenuation storage in three main areas:

    a. Onchan system — oversize pipes on the hill from Onchan to Port Jack
    b. North Douglas system — oversized pipes laid in the flat section of the promenade upstream of the proposed pumping station, supplemented by oversized pipes on the promenade that can also carry the outfall from the Onchan/North Douglas Pump Station

5. The levels of protection to be systematically increased by the combined effort of three courses of action are

    a. Mobilize storage within the system
    b. Apply storm water management techniques to new developments
    c. Intercept storm connections to the foul sewer and hence ultimately reach the objective of reducing overflow frequency to contain a 20-year storm

The strategy enabling the flow of foul water to be intercepted at 3 x DWF with no loss in performance of the existing system, and, over time, under a controlled expenditure profile, may systematically improve performance.

11.8 THE USE OF ATTENUATION STORAGE AND SOURCE CONTROL

To illustrate the practical implications of the strategy, consider one catchment area, the fully developed center of Douglas that drains directly to the short sea outfall at Conister. This area contains both combined and separate sewerage systems, and flow gauging has shown that some 81 ha of the more than 3-km^2 catchment are directly connected to the foul sewer.

At present, the peak flow entering the sewers on a 6-month return period storm is of the order of 1290 L/sec. As the outfall capacity is approximately 700 L/sec it is clear that the stormwater overflows operate many times a year. In the first phase of the strategy it is proposed to intercept the foul flows at a flow rate of 273 L/sec while providing 4700 m^3 of compensation storage to maintain the present frequency of overflow. The next phase of work will utilize the strategy of source control to intercept directly connected surface water to the natural outlets and overland flow routes so as to reduce the frequency overflow from the system to twice a year. This will involve a roll-over program of works to intercept only 28% of the presently connected surface area.

To further increase the level of service provided by the system, a second program of mobilization of system storage will aim to progressively eliminate overflows on the 2-year storm by mobilizing some 740 m^3 of in-pipe storage. This will continue as needed and to contain a 5-year storm will require some 11,000 m^3. The mobilization of these volumes will require some 30 to 35% of the presently available system storage by the installation of flow controls at key points in the system and without the need for any new construction.

In addition, any new development or redevelopment in the area will be under planning control so as to limit storm runoff to predevelopment flows. These fairly unspectacular measures will enable the system to be progressively and economically uprated without the need for extensive construction work and the associated disruption, and will ensure that the system can cope with the projected demands of the future.

11.9 CONCLUSIONS

In areas such as Douglas and Onchan, there are a number of options and proposals to consider in some depth. On one hand, traditional solutions are available, which in whole or in part are likely to be expensive and disruptive. More modern technology has been developed in response to existing and future pollution prevention legislation. In many cases this technology represents superior performance at reduced cost and disruption. Developments are proceeding rapidly in this particular field and it is believed that use of new technology in this case will have inestimable benefits for the Government and people of the Isle of Man, especially with the need to avoid the holiday season for constructional purposes.

The investigations carried out show that the optimum and most cost-beneficial sewerage scheme for the area of Douglas and Onchan is a combination of the various strategies outlined above as follows:

- to provide attenuation storage within the system to reduce pump flows to treatment and the frequency and volume of overflows
- to retain stormwater overflows but replace any existing structures so as to provide partial treatment prior to discharge
- to implement a stormwater management policy for the area so as to reduce the impact of stormwater on the foul drainage system

The lessons to be learned from the development of this plan include:

- the importance of considering all available alternatives early in the planning process and before any final decisions are made

- the identification of the optimum scheme, often achievable by responding to environmental standards that are perceived, but may not be required by legislation
- economic design for the upgrading of facilities usually depending on fully using all existing facilities before embarking on new work
- most importantly, effective performance that is more likely to be achieved by the use of proven "low-tech" and low maintenance equipment that will "fail safe" if anything goes wrong

The use of such strategies has only become feasible and effective in the last few years due to two main developments:

1. the use of computers, which now enable designs to be performed in sufficient detail for the optimum strategies to be found
2. the use of Hydro-Brake™ flow controls that can be relied upon to operate effectively and trouble free

ACKNOWLEDGMENTS

The Authors are grateful to the following people for their permission to present this paper: Mr. N. R. Cooil, Chief Executive, Department of Highways Ports and Properties, Isle of Man Government; Mr. A. J. Newton, Borough Technical Officer, Borough of Douglas and Mr. R. A. Brown, Surveyor, Onchan District Commissioners.

The views expressed in this paper are those of the authors and not necessarily those of the Authority.

REFERENCES

1. Manx Laws and E.C. Directives.
2. Smisson, R.P.M. "A review of the stormwater drainage of new developments," *I.P.H.E.* 8 (1): 64 (1979).
3. Mance, G., and M.M.I. Harman. "The quality of urban stormwater runoff," in *Urban Storm Drainage*, P.R. Helliwell, Ed. (Pentech Press, 1978), p. 603.
4. Fletcher, I.J., C.J. Pratt and G.E.P. Elliot. "An assessment ofthe importance of roadside gulley pots in determining the quality of stormwater runoff," in *Urban Storm Drainage*, P.R. Helliwell Ed. (Pentech Press, 1978), p. 526.
5. Lindholm, O., and P. Balmer. "Pollution in storm runoff and combined sewer overflows," in *Urban Storm Drainage*, P.R. Helliwell, Ed. (Pentech Press, 1978), p. 575.
6. Theil, P.E. "High level of flood protection at low cost," APWA Int. Public Works Congress, Boston (1978).

7. Lumbers, J., R.P.M. Smisson, G.W. Fagan and R. Hudson. "Guidelines for the design of attenuation storage systems," *CONFLOW 88* (1988).
8. Theil, P.E. "New methods of stormwater management," Paul Theil Associates Limited, Bramalea, Ontario (1979).

12 BEST MANAGEMENT PRACTICES FOR URBAN STORMWATER RUNOFF CONTROL

12.1 INTRODUCTION

Urban stormwater runoff has been recognized as one of the major nonpoint sources of pollution, contributing to the degradation of water quality in receiving-water bodies. Characteristics of urban storm runoff and its impact have been well documented.[1] However, urban stormwater pollution control measures are still in the early stages of implementation and relatively few performance data are available from full-scale field applications. Such performance data are needed for the derivation of design criteria for the various structural control measures or "Best Management Practices" (BMPs).

The objective of this paper is to provide a synthesis review of recent advances in the application of structural stormwater management practices. Emphasis will be given to pollutant removal mechanisms and efficiencies of the various BMPs. The concept of an "integrated approach" to planning and designing of urban stormwater quantity and quality control has been gaining global attention. The information presented in this paper should be useful to engineers and planners in dealing with urban stormwater quantity and quality management.

12.2 DETENTION PONDS

12.2.1 General Features

The concept of using stormwater detention basins to reduce runoff pollution gained widespread attention as a result of studies authorized under Section 208

0-87371-805-4/93/$0.00+$.50
© 1993 by Lewis Publishers, Inc.

of the 1972 Clean Water Act established by the U.S. Congress. The "dual-purpose" detention pond design approach allows the pond to (a) reduce flood damages downstream, and (b) to reduce nonpoint pollution from storm runoff.[2] The EPA Nationwide Urban Runoff Project further demonstrated the water quality benefits of wet detention basins.[3]

Dry Ponds

Dry ponds are depressed areas which store runoff during storm events. They are usually designed to reduce the peak flow resulting from a selected design storm (e.g., a 10-year storm) to the predevelopment level to prevent downstream flooding. However, dry ponds are not very effective in removing pollutants; they are basically designed for controlling quantity, not quality. Because of the short detention times, many particulate pollutants do not have enough time to settle out of the runoff, and the ones that do settle to the bottom of the pond are very easily resuspended by the next storm event. Pollutant removal efficiency for dry ponds reported in the literature ranged from 0 to 20% for all pollutants as an average.

Extended Dry Ponds

The outlet structure of a dry pond can be modified in such a way that a "retention outlet" is provided that is sized for slow release of the runoff from a designated "BMP storm". A BMP storm is a small and frequent storm, such as the 1-year storm, which is prescribed by regulations or ordinances as the BMP design storm.

The pollutant removal efficiency for extended dry ponds depends on how long and how much runoff is detained. In general, moderate to high removal rate (40 to 70%) can be achieved for particulate pollutants, such as suspended solids. For dissolved pollutants, such as nutrients, the removal efficiency is very low.

Wet Ponds

Wet ponds, by maintaining a permanent pool, allow particulate pollutants to settle out and dissolved pollutants to be removed by biological uptake or other decay processes. For example, long-term average removal estimates by Driscoll range from around 50 to >90% for total suspended solids (TSS),[3] 40 to 60% for nutrients, and 40 to 45% for zinc. Moderate to high removals for wet ponds were also reported for studies in Florida,[4] North Carolina,[5] and Virginia.[6]

12.2.2 Pollutant Removal Mechanisms

Pollutants are removed in a detention pond mainly through these mechanisms.

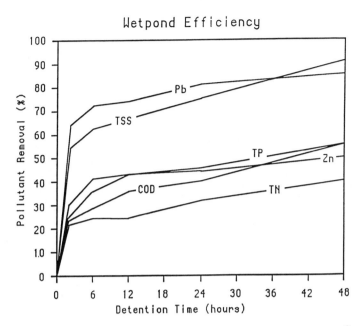

Figure 1.   Removal rate vs. detention time for selected pollutants.[8]

Particle Settling

Particulate pollutants are removed by gravitational settling. Therefore, the removal efficiency for particulates should relate to the inflow particle size distribution of the pollutant and the detention time, which is affected mainly by the size of the pond, the overflow rate, which is the ratio of outflow rate and the pond surface area, the pond geometry, and the design of the outlet structure.

Decay

For nonconservative pollutants such as BOD and pathogenic bacteria, biodegradation and die-off will occur, respectively. Dissolved nutrients are primarily removed by biological activities of the aquatic vegetation in the pond.

For most detention ponds, the dominant factors influencing the removal efficiency are the settling velocity of the pollutants and the overflow rate.

The settlability of various pollutants differ from one another. For example, Whipple and Hunter performed column settling tests and found that hydrocarbons and lead settle out similarly to suspended solids (TSS),[7] but phosphorus, zinc, copper, nickel, and BOD exhibit quite different settling patterns.

Schueler compiled results from a more complete laboratory column test and presented results relating removal efficiency to detention time for a number of pollutants as shown in Figure 1.[8]

The same trend has also been observed in field studies conducted by others, for example, Wu et al.[5] and Yu et al.[6]

12.2.3 Design Considerations

In general, design of a detention pond based on particle settling should be made with the following understandings:

1. Particle settling velocity distribution in the inflow water is a very important design consideration, which is quite site-specific and may vary from storm to storm.
2. Suspended sediments, lead, and hydrocarbons may exhibit similar settling characteristics, whereas phosphorous, nitrogen, and zinc may be grouped into another category.
3. Overflow rate is an important design parameter, which is related to pond size and auxiliary devices, such as baffles, etc.

Generally, the kinetic processes for decay and biological uptake by plants are both enhanced by longer detention time in a pond. Therefore, the detention time can be considered as the key design factor. Longer detention times (24 to 36 h) may be preferred if biological uptake is desired.

12.2.4 Example Design Guidelines

Extended Dry Ponds

Schueler recommended some guidelines for designing BMPs.[8] Highlights of the design considerations for extended dry ponds are:

*   Volume should store runoff quantity produced by one 25-mm storm.
*   For optional pollutant removal, 24 h of detention is desirable.
*   Smaller storms (2.5 to 5.0 mm of runoff) should be detained for at least 6 h.
*   A two-stage design is recommended, an "upper" stage of the pond is to remain dry normally, and a "bottom" stage is regularly inundated with its volume set to store about 15 mm of runoff.
*   Marshes should be established at the bottom stage.
*   The outlet control device should be designed to set water levels and should withstand partial clogging.
*   A low-flow channel is desirable.

An example schematic diagram of a design for an extended dry pond is shown in Figure 2.

Wet Ponds

By maintaining a permanent pool, wet ponds achieve particulate and dissolved pollutant removal through enhanced particle settling, decay processes,

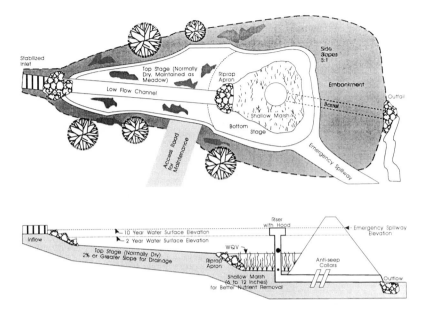

Figure 2.   Schematic of extended detention pond design features.[8]

and biological uptake. In addition to the particulate settling-based design approach, biological and other decay processes should also be included in deriving design guidelines. In general, wet pond design methodology could include the following approaches:

1. *Solids Settling Design:* Based on sedimentation theory, the method uses the particle settling velocity and overflow rate as key parameters. Pond size and configuration are designed so that particle settling is optimized. Figure 3 illustrates a schematic diagram of a typical wet pond.[8]

2. *Lake Eutrophication Model Design Method:* Hartigan proposed that a wet pond be considered as a small eutrophic lake that can be simulated by empirical models to evaluate lake eutrophication.[9] Hartigan used the "input-output" phosphorus retention model developed by Walker as the design tool.[10] The Walker model relates phosphorus removal to such variables as the inflow total phosphorus concentration, the second order decay rate, mean lake depth, and the hydraulic residence time. By changing the wet pond volume and other geometry characteristic, one can obtain the removal efficiency desired.

3. *Detailed Hydraulics/Water Quality Modeling Approach:* A wet pond can also be modeled to a more detailed fashion, analogous to a lake. Flow patterns, pollutant transport, and transformation processes in a pond can be simulated under a variety of trial design conditions so that some guidelines can be obtained. For example, the geometry of the pond can be changed, or a baffle installed, and their effect on the removal efficiency can then be examined by using a model.[11]

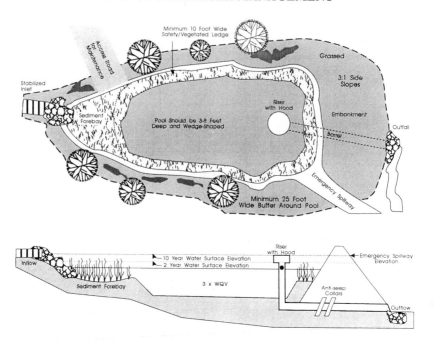

Figure 3.   Schematic diagram of a wet pond.[8]

A summary of some design recommendations for wet pond, found in the literature is given in Table 1.

12.3 INFILTRATION FACILITIES

12.3.1 General Features

Infiltration Trenches and Basins

These facilities are trenches or basins in which coarse sand or gravel is placed. Filter fabric can be used to line the trench or basin to prevent pollutants from entering the groundwater.

Porous Pavement

Porous pavement generally consists of a thin layer of open-graded asphalt mix on top of a deep base filled with large-size crushed stone aggregate to form a "reservoir" for detaining stormwater. A filter fabric may also be installed to protect groundwater.

Moderate to high pollutant removal efficiencies can be expected of properly designed and maintained infiltration trenches and basins. For example, Schueler

Table 1. Summary of Wet Pond Design Recommendations

Design Parameter	Recommended Design
1. Storage volume (permanent pool)	Volume of basin at least 2.5 times mean storm runoff volume.
2. Depth (permanent pool)	Average 1 to 3 m. Use large depth if practical.
3. Side slopes	No steeper than 3H:1V.
4. Length/width ratio	At least 3:1.
5. Baffles	Use as needed. Should maximize the flow length.
6. Vegetation	Marsh establishment near inlet and perimeter.
7. Sediment forebay	Shallow forebay for sedimentation and vegetation is preferred.

reported removal efficiencies of 80 to 100% for suspended sediment, 30 to 70% for nutrients, and 15 to 80% for metals.[8]

Information on water quality benefits of porous pavements are limited. However, significant pollutant loading reductions have been observed by, e.g., Pratt et al.[12] and Hogland et al.[13]

12.3.2 Pollutant Removal Mechanisms

Infiltration trenches and basins, and porous pavements function in a similar fashion. These practices allow stormwater runoff to filter through the soil column where pollutant removal by physical (sedimentation, adsorption, etc.), chemical (reaction), and biological (root uptake, transformation, etc.) processes take place. Infiltration facilities can achieve fairly high to high degree of removal of particulates as well as dissolved pollutants, if properly designed. Since clogging by sediment is a major concern for infiltration facilities, some type of pretreatment device, such as a vegetative buffer strip, is usually included as part of the infiltration system.

12.3.3 Design Considerations

A typical cross-section of an infiltration trench is shown in Figure 4. The size of the trench is determined by the amount of storage volume required. Sizing rules for storing either 12 mm or 25 mm of runoff per impervious area in the

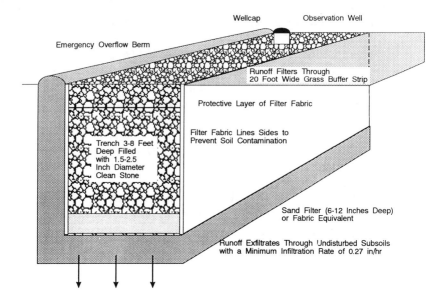

Figure 4.   A typical cross-section of an infiltration trench with monitoring well.[14]

contributing watershed have been recommended.[8] The trench depth is usually between 1 and 3 m. Filter fabric is placed around all sides of the trench to prevent clogging by soil fines. Buffer strips should be placed between runoff producing area and the trench for solids removal.

A typical porous pavement cross section is shown in Figure 5.[15] The depth of the stone reservoir should be designed so that, as a minimum, the first 15 mm of runoff is detained for no longer than 72 h, or the average time interval between storm events. Underground drains to a holding pond may be needed for soils with low permeability.

Due to concerns regarding the durability of porous pavements and the complexity involving construction specifications, porous pavements as a BMP are usually recommended for low-traffic roads or parking lots. More studies are needed so that water quality benefits and the durability of porous pavements can be fully documented.

12.4 VEGETATIVE FILTER STRIPS

12.4.1 General Features

Vegetative (usually grass, sometimes wood) filter strips are low-cost practices that have been found to offer some water quality benefits. A "level spreader" is usually needed as a component of a filter strip for the purpose of

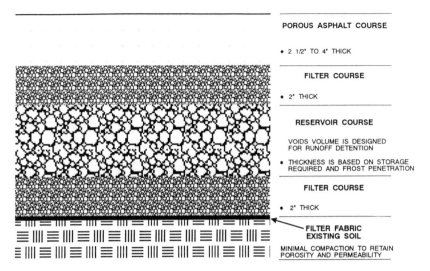

PParser POROUS ASPHALT COURSE

- 2 1/2" TO 4" THICK

FILTER COURSE

- 2" THICK

RESERVOIR COURSE

VOIDS VOLUME IS DESIGNED
FOR RUNOFF DETENTION

- THICKNESS IS BASED ON STORAGE
REQUIRED AND FROST PENETRATION

FILTER COURSE

- 2" THICK

FILTER FABRIC
EXISTING SOIL

MINIMAL COMPACTION TO RETAIN
POROSITY AND PERMEABILITY

Figure 5.   Typical porous pavement section.[15]

spreading stormwater runoff evenly onto the strip. Otherwise channels may form and the strip will be "shortcircuited" and lose its removal efficiency. Figure 6 shows the plan view of a level spreader/vegetative filter strip system tested in Charlottesville, Virginia.[6] The system shown in Figure 6 was found to be fairly effective in removing particulate pollutants. Details of the field test of the level spreader/vegetative filter strip system are presented elsewhere.[6]

Vegetative filter strips can be used as a "first stage" practice, preceding another practice, so that a high overall performance is achieved. For example, runoff from a parking lot can be made to pass over a filter strip before entering an infiltration trench. Not only will the combined removal be higher, but also the infiltration trench will be less likely to be clogged by particles.

12.4.2 Pollutant Removal Mechanism

Figure 7 depicts a level spreader/vegetative filter strip installation. The system essentially functions as two best management practices (BMPs), namely, a minidetention pond when runoff is retained in the level spreader, and a vegetative filter strip when runoff spills over the weir and onto the grass strip. Although usually having a small volume and depth (3 ft or 1 m in this case), the level spreader does not act as a flow-through basin until it overflows. It is therefore expected that a fair amount of pollutants will be trapped at the bottom of the level spreader due to settling.

The vegetative filter strip, or VFS, serves to slow down overland flow, allowing sediments and pollutants to settle out or infiltrate. Mechanisms associated with pollutant removal for such a system include:

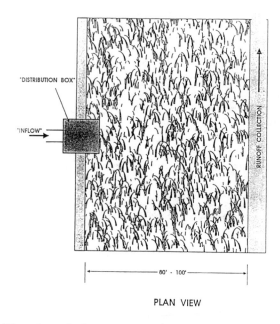

PLAN VIEW

Figure 6. Plan view of a level spreader/vegetative filter strip system.

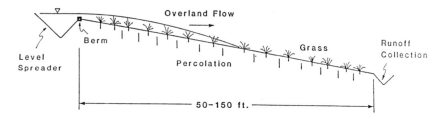

Figure 7. Profile of a level spreader/vegetative filter strip.[6]

- sedimentation and filtration, removing primarily solids and metals
- adsorption, precipitation and plant uptake, removing primarily nutrients

12.4.3 Design Considerations

The length of the filter strip is an important design parameter. A filter length of at least 20 m is desirable. However, other factors, such as slope, runoff velocity, particles' size distribution, and flow depth are all significant factors in determining the overall pollutant removal efficiency of a filter strip.

Wong and McCuen developed a nomograph for sizing the filter length for given slope, runoff velocity, and desired solids removal efficiency.[16] Such charts are very useful, especially with additional pollutants, such as nutrients and metals, included when data are available.

12.5 WETLANDS

12.5.1 General Features

Wetlands have long been used for final treatment of municipal wastewater. Many studies have demonstrated the cost-effectiveness of wetlands.

Utilization of wetlands for treatment of urban runoff is a recent idea. A study conducted in Minnesota by Barten examined the use of natural marshes to filter nutrient-rich urban runoff water.[17] Significant removal of nutrients and suspended solids was observed. However, very little information is available regarding the cost-effectiveness of using natural or artificial marshes as a BMP.

A study conducted by the U.S. Geological Survey in cooperation with the Florida Department of Transportation was intended to examine the efficiency of a detention pond/wetland system in treating highway runoff. The system, described by Martin, receives runoff from a four-lane highway and the adjacent residential area.[4] Stormwater runoff enters the detention pond first, and then the wetland, before discharging into the receiving water.

Martin reported that the wetlands system was quite efficient in removing metals, i.e., lead, 73%; zinc, 56%; and suspended solids, 66%.[4] Results for nutrients were low, ranging from 17% for total phosphorous to 21% for total nitrogen.

Another study conducted by Scherger and Davis of a natural wetland in Michigan showed good removal efficiency for solids (76 to 93%), moderate for total phosphorous (40 to 60%), and low for nitrogen (20 to 30%).[18]

12.5.2 Pollutant Removal Mechanism

Wetlands are complex ecosystems characterized by high floral productivity and nutrient needs, high decomposition rates, low oxygen content in the sediments and substrates, and large adsorptive surfaces in the substrates.[19]

Removal mechanisms of wetlands include physical processes, such as sedimentation for particulate pollutants; adsorption for ammonium ions, phosphate, metals and viruses; chemical precipitation for metals; filtration for organic matter, phosphorus, bacteria, and solids; volatilization for oils, chlorinated hydrocarbons, and mercury; and biological processes, such as nutrient uptake.[20]

12.5.3 Design Considerations

Very little scientific information is available regarding design criteria for wetlands used for stormwater treatment. The limited literature findings suggest that important factors to be considered in wetland design and management are pretreatment (for example, a detention pond before a wetland) for solids removal and wetland hydrology.[20]

12.6 MINING PONDS IN MALAYSIA

As in the U.S., the management of stormwater quantity and quality under rapid urban growth has been gaining public attention in Malaysia in recent years. A case example is described in the following sections.

12.6.1 Use of Mining Ponds in the Klang Valley

Stormwater flooding and polluted discharges due to concentrated human activities have become major issues in many urban centers in the Klang River Basin, Malaysia. The problems are even more severe with the occurrence of frequent intense rainfalls; the geomophorlogical nature of the basin; the inadequacy of present drainage systems; the patterns of urbanization, especially over water ponding areas, such as mining ponds; the relatively poor waste management practices; and the lack of stormwater and pollution control facilities.

Historically the Klang Valley was the major center for the world's tin production, and in this area there now exist more than 80 abandoned mining ponds located primarily in the Jinjang, Batu, and Kerayong subwatersheds. In the past, these ponds have been filled for use in urban development. However, this practice has now given way to a consideration of their potential application for recreation use and runoff pollution reduction. Although the concept of using mining ponds for flood detention was first recommended by the Kuala Lumpur Drainage Master Plan 15 years ago, such implementation has been slow with emphasis placed on the improvement of trunk drainage rather than the promotion of stormwater control.

Mining ponds found in the Klang River Basin differ quite substantially in size with ponds ranging from 0.3 ha to 90 ha, a maximum depth of 30 m, and a total area of some 800 ha. Despite the existence of many ponds, only those with superior storage capacity and those that are also located close to rivers would be of potential use for runoff and pollution reduction. A study of four subwatersheds in Kuala Lumpur indicated that pond sizes of 1 to 2% of the upstream watershed area are sufficient to regulate the runoff from a 100-year storm.[21]

A simulation study has shown that flood discharges of the Jinjang, Keroh, Bunus, Kerayong, and Damansara rivers were tremendously attenuated using different combination of ponds within each subwatershed area.[22] For example, through the use of two ponds, the Bunis River can reduce the flow of its 100-year flood from 108 m^3/sec to 50 m^3/sec. Several ponds are now being used for controlling stormwater floodings, namely at Jinjang (four ponds in series), Segambut, and Ampang Hilir. Study of the Ampang Hilir pond concluded that substantial runoff amounts from the Kampong Pandan area can be temporarily

Table 2. Jinjang Ponds Water Quality Sampling (July 23, 1988)

Parameters	Inflow	Outflow	% Removal
COD (mg/L)	67	20.4	69
BOD (mg/L)	14	3.7	73
SS (mg/L)	382	10.0	97
Turbidity (JTU)	158	7.1	95

stored (with a maximum water level rise of 15 cm) resulting in an acceptable outflow discharge to the previously flooded downstream Armco drain at Jin, Tun Razak.[23]

12.6.2 Water Quality Benefits of Mining Ponds

Mining ponds can also serve as stilling basins for water quality control, depositing silt and sand, and even trapping floating debris. Initial field investigations of pollutant removal at the Jinjang ponds were carried out in 1988. A tremendous improvement in the overall water quality was observed (Table 2).

These and future control ponds have been earmarked for eventual use as urban ponds/lakes. The Titiwangsa (8.7 ha), Ampang (3.1 ha), Perdana (1.9 ha), and Shah Alam (3.0 ha) ponds are now used entirely for recreational purposes such as fishing, windsurfing, and boating. The Perdana pond, however, still regulates a relatively small amount of the incoming flows to the Klang River.

There is currently great interest in developing the Jelatek mining pond as an urban pond. A bathymetric survey was done in 1989 that indicated at normal water level, storage of $170 \times 10^3 m^3$ and a surface area of $22 \times 10^3 m^2$. Although it is now in the design stage, the Jelatek pond will reduce the flow from Taman Setiawangsa by at least 50 m^3/sec for a 100-year storm. Beautification and landscaping of the pond corridor for eventual recreational purposes will come at a later stage, when a satisfactory water quality standard has been met.

The year 1991 seems to have marked the beginning of government interest in embarking on a more engineered method of developing mining ponds for stormwater quantity/quality control and recreation. Studies are currently being carried out to establish design criteria and standards for such implementation in the tropical urban environment.

In summary, the use of mining ponds have contributed significantly to the control of stormwater flooding, as well as water quality improvement, in some rivers in the Klang Valley. Those ponds that were converted into urban ponds/lakes have constituted a valuable resource for sustaining a range of ecosystems and water bodies of considerable scientific and recreational value. A visionary and properly planned urbanization in the Klang River Basin must incorporate the remaining strategic ponds for stormwater and pollution control and recreational benefits.

REFERENCES

1. "Results of the Nationwide Urban Runoff Program, Vol. I. Final Report," U.S. EPA, Water Planning Division, Washington, D.C. NTIS No. PB84-185552 (1983).
2. Whipple, W., Jr., R. Knapp and S. Burke. "Implementing dual-purpose stormwater detention programs," *J. Water Res. Plan. Manage. ASCE* 113 (6):779 (1987).
3. Driscoll, E.D. "Methodology for analysis of detention basins for control of urban runoff quality," U.S. EPA Report-440/5-87-001 (1986).
4. Martin, E.H. "Effectiveness of an urban runoff detention pond-wetlands system," *J. Environ. Eng. Div. ASCE* (1988).
5. Wu, J.S., B. Holman and J. Dorney, "Water quality study on urban wet detention ponds," Proceedings of the Engineering Foundation Conference on Current Practice and Design Criteria for Urban Quality Control, ASCE, Potosi, MO (1989), p. 280.
6. Yu, S.L., W.K. Norris and D.C. Wyant. "Urban BMP demonstration project in the Albemarle/Charlottesville area," Final Report to Virginia Dept. of Conservation and Historic Resources, University of Virginia, Charlottesville, VA (1987).
7. Whipple, W., Jr., and J.V. Hunter. "Settleability of urban runoff pollution," *J. Water Pollut. Control Fed.* 53:1726 (1981).
8. Schueler, T.R. "Controlling urban runoff: a practical manual for planning and design urban BMPs," Metropolitan Washington Council of Governments, Washington, D.C. (1987).
9. Hartigan, J.P. "Basis for design of wet detention basin BMPs," Proceedings of the Engineering Foundation Conference on Design of Urban Runoff Quality Controls, ASCE, Potosi, MO (1988), p. 122.
10. Walker, W.W., Jr. "Phosphorus removal by urban runoff detention basins," Lake and Reservoir Management Society, Washington, D.C. (1987).
11. Yu, S.L., Y. Wu and J. Benelmouffok. "Modeling the effect of baffles on the hydrodymanic and pollutant removal characteristics of a detention pond," Report to USGS, University of Virginia, Charlottesville, VA (1989).
12. Pratt, C.J., J.D.G. Manele and P.A. Schofield. "Porous pavements for flow and pollutant discharge control," Proceedings of the Fifth International Conference on Urban Storm Drainage, Osaka, Japan (1990) p. 839.
13. Hoglund, W., M. Larson and R. Berndtsson. "The pollutant build-up in pervious road construction," Proceedings of the Fifth International Conference on Urban Storm Drainage, Osaka, Japan (1990), p. 845.
14. Harrington, B.W. "Design and construction of infiltration trenches," Proceedings of the Engineering Foundation Conference on Design of Urban Runoff Quality Controls, ASCE, Potosi, MO (1988), p. 122.

15. Diniz, E.V. "Porous Pavement: Phase 1 — Design and operational criteria," U.S. EPA Report-600/2-80-135.

16. Wong, S.L., and R.H. McCuen. "The design of vegetative buffer strips for runoff and sediment control in stormwater management in coastal areas," Maryland Department of Natural Resources, Annapolis, MD (1982).

17. Barten, J.M. "1981 Clear Lake Water Quality and Treatment Marsh Assessment," U.S. EPA Report-S804691-01-0, Washington D.C. (1982).

18. Scherger, D.A., and J.A. Davis. "Control of stormwater runoff pollutant loads by a wetland and retention basin," in Proceedings of the International Symposium on Urban Hydrology, Hydraulics and Sediment Control, Univ. of Kentucky, Lexington, KY (1982).

19. Dorman, M.D., J. Hartigan, F. Johnson and B. Maestri. "Retention, detention and overland flow for pollutant removal from highway stormwater runoff: interim guidelines for management measures," U.S. FHWA Report-FHWA/RD-87-056, Federal Highway Administration (1988).

20. Livingston, G.H. "The case of wetlands for urban stormwater management," Proceedings of the Engineering Foundation Conference on Design of Urban Runoff Quality Controls, Potosi, MO (1988), 467.

21. "Study on flood mitigation of the Klang River basin," JICA Final report, submitted to the Government of Malaysia, 1989.

22. Talib, S.O. "Siting of ex-mining ponds for stormwater and NPS pollution control in the Klang River basin," Undergraduate Thesis, UTM, Johor, Malaysia (March 1990), in Malay.

13 MWRA CSO CONTROL PROGRAM — ONLY PART OF THE SOLUTION

13.1 INTRODUCTION

The Massachusetts Water Resources Authority (MWRA) is a regional water and sewerage authority which provides wholesale water and sewer services to the Metropolitan Boston area. The MWRA was created by the Massachusetts Legislature as an independent authority in 1985 following the filing of a lawsuit in federal court to correct the pollution of Boston Harbor resulting from inadequate and poorly maintained sewerage facilities. Immediately after its creation, the MWRA began to tackle treatment plant issues, including addressing the need for a secondary treatment plant, ceasing the ocean discharges of sludge, and making interim repairs to the existing, antiquated treatment plants. The MWRA's initial activities led to a court-ordered schedule for construction of a single new primary and secondary treatment facility, development of sludge management plan focused principally on reuse, and completion of an accelerated program of repairs to existing facilities.

Once this schedule had been negotiated, the U.S. Environmental Protection Agency (EPA), a plaintiff in the federal court case, asked that the MWRA take "responsibility" for controlling discharges from combined sewer overflows (CSOs) within its service area. Combined sewers (sewers where sanitary sewage and stormwater runoff are carried in a single pipe) currently discharge over 10 billion gal (38,000 ML) of untreated wastewater into Boston Harbor and its tributaries annually, and EPA recognized this as yet another significant cause of pollution in the Harbor. Even though most of these CSOs are relief points from community systems rather than from the MWRA system, the MWRA agreed to the EPA's request to take responsibility because the MWRA agreed that the problem of pollution from CSOs had to be addressed as part of the general effort to revitalize the water quality of Boston Harbor.

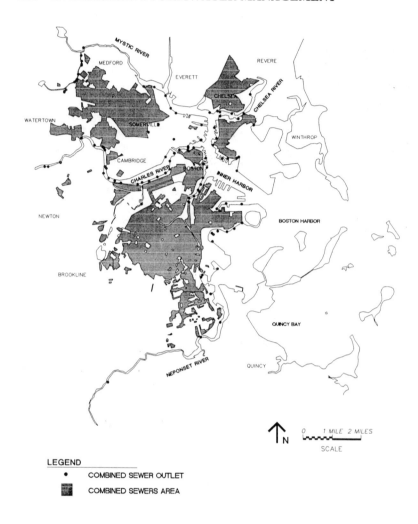

CSO FACILITIES PLAN/*CEM* HILL TEAM

Figure 1.   Combined sewer area and outlets.

13.2 THE CSO PROBLEM IN THE MWRA SERVICE AREA

Combined sewers and combined sewer overflows are found on four communities in the metropolitan Boston area: Boston, Cambridge, Chelsea, and Somerville. Figure 1 shows the combined sewer area of about 12,000 acres (4900 ha). These communities discharge combined sewage into the MWRA's interceptor system. There are some additional combined sewer areas within the MWRA's service area, for example, in the Town of Brookline. However, these systems no

Table 1. Annual CSO Volumes[a]

Location	Existing Conditions	No Further CSO Action[b]	Percent Reduction
Alewife/Mystic	228	71	68
Charles River Basin[c]	4037	1975	51
Boston Inner Harbor	6400	2512	61
Dorchester Bay	859	438	49
Neponset River Estuary	39	20	49
Total	11536	5016	57

[a]   All volumes in millions of gallons.
[b]   "No Further CSO Action" refers to conditions once improvements to the Deer Island WWTF are completed but no further CSO control has been completed.
[c]   Includes upper and lower Charles River Basin.

longer have functioning CSO discharge points, and these systems have not been included in the MWRA's CSO planning program.

Together, the combined sewer communities and the MWRA have a total of approximately 85 permitted CSOs. The CSOs are distributed in six drainage basins: Boston's Inner Harbor, Dorchester Bay, the lower Charles River Basin, the upper Charles River Basin, the Neponset River, and the Alewife Brook/ Mystic River. Three basins, the Inner Harbor, Dorchester Bay and the lower Charles River, receive the vast majority of the combined sewage discharged. Table 1 shows the volumes of combined sewage discharged into each basin in the combined sewer area. Discharges into the Neponset River and the Alewife Brook account for less than 3% of the total volume of combined sewage, and the upper Charles River Basin has only negligible CSO discharges. On average, one or more of the CSOs in the six basins discharge 60 to 80 times per year, or almost every time it rains. Table 2 shows the number of discharge events for each basin.

With existing treatment capacity, the CSO discharges into the six basins release a total of approximately 10 to 12 billion gal (38,000 to 45,000 ML) of combined sewage into Boston Harbor every year, about 5% of the total waste-water flows in the system. Upgrades to the MWRA system, principally the improvement of pumping capacity at the Deer Island Wastewater Treatment Facility (WWTF), will improve the transport capacity to the WWTF. Once these improvements are completed, as shown on Table 1, the volume will be reduced by more than half to just over 5 billion gal (19,000 ML) annually. As can be seen from the data on Table 2, while there will be some reduction in the number of discharges, it is relatively small. Table 3 shows annual pollutant loadings from CSOs and the reductions which occur after improvements to the WWTF.

Table 2. Number of CSO Overflow Events Annually

Location	Existing Conditions	No Further CSO Action[a]	Percent Reduction
Alewife/Mystic	40	22	45
Upper Charles Basin	3	1	67
Lower Charles Basin	60	47	19
Boston Inner Harbor	80	68	15
Dorchester Bay	70	66	6
Neponset River Estuary	43	43	0

[a] "No Further CSO Action" refers to conditions once improvements to the Deer Island WWTF are completed but no further CSO control has been completed.

Table 3. Annual CSO Pollutant Reductions

	Total Suspended Solids[a]		
Location	Existing Conditions	No Further CSO Action[b]	Percent Reduction
Alewife/Mystic	212	66	69
Charles River Basin[c]	6494	3175	51
Boston Inner Harbor	13163	5158	61
Dorchester Bay	951	485	49
Neponset River Estuary	33	17	48
Total	20853	8903	57
	BOD[a]		
Alewife/Mystic	123	38	69
Charles River Basin[c]	2981	1459	51
Boston Inner Harbor	5844	2290	61
Dorchester Bay	497	253	49
Neponset River Estuary	20	10	50
Total	9465	4050	57

[a] TSS and BOD in 1000 lb.
[b] "No Further CSO Action" refers to conditions once improvements to the Deer Island WWTF are completed but no further CSO control has been completed.
[c] Includes upper and lower Charles River Basin.

Even with these reductions in CSO discharges and the resulting reductions in pollutant loadings, however, the pollution attributable to CSO discharges would be considerable. There would still be some 50 to 70 CSO discharges each year, causing elevated bacterial concentrations and contributing substantial solids and toxics to the system. Although the MWRA's improvements to the WWTF would, and already have, contributed to improvement to surface water quality, CSOs would continue to be a significant source of pollution. Thus, both the MWRA and U.S. EPA were well aware that further CSO control needed to be studied.

13.3 MASSACHUSETTS SURFACE WATER QUALITY STANDARDS

In any CSO facilities planning program, planners are faced with the very difficult question of how much CSO control will be enough. CSO discharges are considered "point sources" by U.S. EPA and require NPDES permits. As such, under U.S. EPA's interpretation of the U.S. Clean Water Act (U.S. CWA, 33 USC 1251), CSO discharges must meet appropriate "technology-based standards" (Sections 301(b)(1)(B) and 301(b)(2)) and "any more stringent limitation" necessary to meet Water Quality Standards (WQS) (Section 301(b)(1)(C)). Water quality monitoring and modeling in Boston Harbor led to an early conclusion that the requirement to meet WQS was the more stringent associated with CSO control, and planning thus focused on control programs that would be consistent with the WQS.

The Massachusetts WQS, which are developed and enforced by the Massachusetts Department of Environmental Protection (DEP), consist of two basic components: designated beneficial uses and water quality criteria, many of which are numeric standards, to protect those beneficial uses. In the U.S., there has been a national goal established to classify all waters as swimmable and fishable. In Massachusetts, the defined water body classifications, which contain the designated beneficial uses, include fresh and salt water categories (C and SC) that are not fishable and swimmable. Table 4 shows the definitions of the water body classifications based upon designated beneficial uses. In developing new standards which were promulgated just as the MWRA completed its CSO facilities plan in September 1990, the DEP determined that all waters would be classified as fishable and swimmable, so the C/SC classification is not presently applied to any waters in Massachusetts.

Previously, the only waters classified as C or SC were the Charles River Basin and Boston Harbor, in large part because of CSO discharges into those waters.[*] Under the WQS which were in effect until September 1990, CSO discharges

---

[*] Although the c/sc category might be viewed as allowing less aggressive CSO control, in fact, the numeric bacterial standards were, and still are, not significantly lower than those for the B/SB category. It would not be possible to have even moderate-size CSO discharges into most water bodies and meet this lower standard.

Table 4. Definition of Water Uses and Classifications for Massachusetts

Classes for Coastal and Marine Waters

Class SA. These waters are designated as an excellent habitat for fish, other aquatic life and wildlife and suitable for primary and secondary contact recreation. In approved areas they shall be suitable for shellfish harvesting without depuration (Open Shellfish Areas). These waters shall have excellent aesthetic value. (314 CMR 4.05[4][a])

Class SB. These waters are designated as a habitat for fish, other aquatic life and wildlife, and for primary and secondary contact recreation. In approved areas they shall be suitable for shellfish harvesting with depuration (Restricted Shellfish Areas). These waters shall have consistently good aesthetic value. (314 CMR 4.05[4][b])

Class SC. These waters are designated as a habitat for fish, other aquatic life and wildlife; and for secondary contact recreation. They shall also be suitable for certain industrial cooling and process uses. These waters shall have good aesthetic value. (314 CMR 4.05[4][c])

Classes for Inland Waters
Class A. These waters are designated as a source of public water supply. To the extent compatible with this use they shall be an excellent habitat for fish, other aquatic life and

wildlife and suitable for primary and secondary contact recreation. These waters are designated for protection under Section 4.03(3) of the regulations. (314 CMR 4.05[3][a])

Class B. These waters are designated as a habitat for fish, other aquatic life and wildlife, and for primary and secondary contact recreation where designated. They shall be acceptable for public water supply with appropriate treatment. They shall be suitable for irrigation and other agricultural uses and for compatible industrial cooling and process uses. These waters shall have consistently good aesthetic value. (314 CMR 4.05[3][b])

Class C. These waters are designated as a habitat for fish, other aquatic life and wildlife; and for secondary contact recreation. These waters shall be suitable for irrigation of crop used for consumption after cooking and for compatible industrial cooling and process uses. These waters shall have good aesthetic value. (314 CMR 4.05[3][c])

Use Related Definitions

Aquatic Life. A native diverse, community of aquatic flora and fauna.

Coastal and Marine Waters. The Atlantic Ocean and all contiguous saline bays, inlets and harbors within the jurisdiction of the Commonwealth including areas where fresh and salt waters mix and tidal effects are evident or any partially enclosed body of water where the tide meets the current of a stream or river.

Cold Water Fishery. Waters in which the maximum mean monthly temperature generally does not exceed 68 °F (20 °C) and when other ecological factors are favorable (such as habitat), are capable of supporting a year-round population of cold water stenothermal aquatic life such as trout (salmonidac).

Inland Waters or Fresh Waters. Any surface water not subject to tidal action or not subject to the mixing of fresh and ocean waters.

Lakes and Ponds. Waterbodies situated in a topographic depression or a dammed river channel with water usually not flowing and an area greater than 20 acres; or less than 20 acres if the water depth in the deepest part of the basin exceeds 2 meters (6.6 feet) or if a discrete shoreline makes up all or part of the boundary. Exceptions include impervious manmade retention basins; river impoundments with flowing water, and harbors and bays which have year round navigable access to the ocean.

National Goal Uses. Propagation of fish, shellfish, other aquatic life and wildlife, and recreation in and on the water in accordance with the Federal Act (Section 101(a)(2)).

Primary Contact Recreation. Any recreation or other water use in which there is prolonged and intimate contact with the water with a significant risk of ingestion of water. These include, but are not limited to, wading, swimming, diving, surfing, and water skiing.

Rivers and Streams. Waterbodies contained within a channel (naturally or artificially created) which periodically or continuously contains flowing water or forms a connecting link between two bodies of standing water.

Secondary Contact Recreation. Any recreation or other water use in which contact with the water is either incidental or accidental. These include but are not limited to fishing, boating, and limited contact incident to shoreline activities.

Warm Water Fishery. Waters in which the maximum mean monthly temperature generally exceeds 68 °F (20 °C) during the summer months and that are not capable of sustaining a year-round population of cold water stenothermal aquatic life.

Source: 314 CMR 4.00 (Mass. DWPC 1990a).

were not specifically addressed. There was a tacit assumption that significant CSO discharges would cause a water body to be classified as C/SC and that this designation would address the issue of CSO discharges.

In developing revisions to the WQS standards, DEP adopted a partial use category which would be assigned to a water body if pollution would *occasionally* degrade the water quality and impair the continuous attainment of a designated use. The policy for implementation of the partial use category was spelled out in the DEP's CSO policy, for which guidance was released with the revised WQS. This policy requires the completion of a CSO facilities plan to demonstrate the cost-effectiveness of various CSO control measures, including separation of combined sewer systems, and sets a goal of no more than four discharges per year where CSO control facilities, rather than sewer separation, are selected.

These measures for regulation of CSO discharges are associated with the first component of the WQS, beneficial use designations. The second component of the WQS, the water quality criteria, also has implications for CSO control. The water quality criteria are either numeric or narrative and are intended to support the designated beneficial uses as defined in the classifications. If these criteria were not exceeded in a water body (outside a defined mixing zone), there would be no impairment of the designated uses, and no requirement for CSO control beyond the technology-based requirements.

The numeric criteria for bacterial pollutants are specified as no more than 200 fecal coliform organisms per 100 ML for Class B/SB waters for the geometric average, with no more than 10% of the samples greater than 400. Any appreciable amount of raw sewage, even when mixed with a large volume of stormwater, will cause an exceedance of that number. While bacterial pollution is the major cause of this impairment, there are other sources of concern, including the potential for elevated levels of various metals and toxics, and the likelihood of aesthetic contamination, which must be controlled in all waters. In sum, CSO discharges *do* cause impairment of beneficial uses and must be controlled consistent with the WQS and the CSO policy.

The recommended facilities, discussed below in Section IV, would control CSO discharges to no more than four per year, consistent with the goals in the DEP CSO policy. The MWRA has applied for partial use designations for each water body that would be subject to occasional CSO discharges, the next step in the Massachusetts regulatory process. While the DEP has not acted on this request, it has indicated that the facilities plan generally complies with the WQS and the CSO policy, and the MWRA anticipates that the designation will be granted. Since there will be a lengthy design and construction period before the CSO control facilities are completed, there is no present need for immediate action by DEP on the request.

13.4 THE CSO FACILITIES PLAN

Upon agreeing to EPA's request to take responsibility for CSOs, the MWRA began work on several small CSO projects identified during earlier studies of the

CSO problem and also initiated a new facilities planning effort to identify long term strategies to achieve CSO control. The facilities planning program, which included environmental review was conducted by a consultant team led by CH2M Hill and completed in September 1990.[1]

The facilities planning program reviewed the technologies available to control CSO discharges and evaluated each for both ability to meet water quality standards and cost-effectiveness. The outcome of the study was driven in large measure by the Massachusetts WQS which prescribe relatively low levels for bacterial contamination. Since the presence of any significant volume of untreated raw sewage in a water body causes the bacterial standards to be exceeded, CSO control must either treat or virtually eliminate CSO discharges. For most parts of the combined system, the study concluded that storage, combined with later pump-back and treatment at the MWRA's Deer Island WWTF, was the most cost-effective solution that would meet water quality requirements. In addition, storage and treatment was found to be the most cost effective technology on a unit cost basis. The cost-effectiveness of the storage and treatment option is not surprising given the large flows at some discharge points. For example, one CSO has a peak flow rate of 1200 million gal/day (4500 ML/day).

The facilities planning program led to a recommended CSO control plan for the combined sewer areas in the four CSO communities. The recommended facilities included a deep tunnel storage system to serve the largest basins in the combined sewer area, a near surface storage facility for the Alewife Brook, and the addition of a limited number of new storm sewers in the Neponset River Basin. Figure 2 shows the proposed deep tunnel storage system and Figure 3 shows the locations of all the proposed facilities. In addition, the study found that prioritization of the storage volume for use in capturing overflows from specific basins could effect further improvements in water quality. Thus, the recommended CSO control plan outlines an operational plan for the deep tunnel system in addition to the recommendation for the control facilities themselves. The recommended facilities and operating plan are consistent with the goals set by DEP in the state's CSO policy.

The deep tunnel storage system will link major basins with a shared storage tunnel. During storm events, diversion structures will redirect existing combined flows into consolidation conduits for transport to drop shafts connecting the surface facilities to the tunnel system. After the end of the storm event, when capacity is available in the MWRA system, two pump stations will discharge stored flows to existing MWRA headworks facilities. At the headworks, the combined sewage will enter the MWRA system and be transported to the Deer Island WWTF for treatment. Following treatment, the effluent will be discharged through an ocean outfall.

The storage tunnels will have a capacity of 315 million gal (1194 ML). The surface facilities and drop shafts will provide an additional 27 million gal (102 ML) of storage capacity, giving the storage system a capacity of 342 million gal (1296 ML). In large storm events, on average about four times per year, the capacity of the storage system would be exceeded and the portion of the

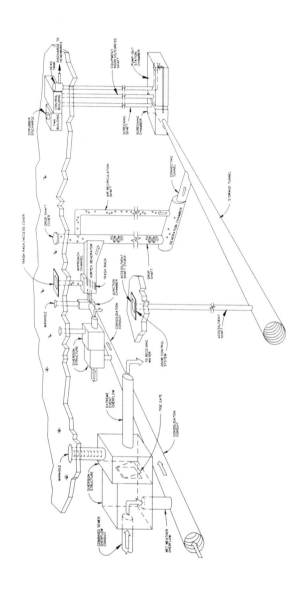

CSO FACILITIES PLAN/CRM*HILL* TEAM

Figure 2.   Schematic of CSO storage tunnel system.

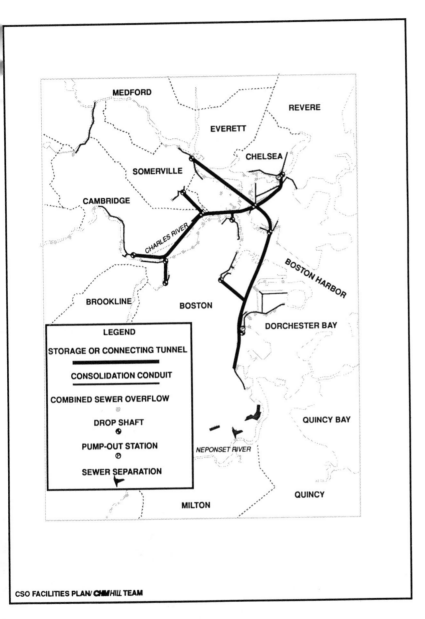

Figure 3.   CSO plan: an overview.

combined sewage that could not be accommodated by the storage system would be discharged from a limited number of remaining CSOs. Screening of these discharges would be provided at some locations, but no other treatment is planned.

The total design and construction cost for all recommended facilities is approximately $1.2 billion, expressed in 1990 dollars. The sewer separation and near-surface storage projects account for less than $100 million. Thus, virtually all of this cost is associated with the deep tunnel storage system.

One of the benefits of the proposed CSO control program is the flexibility it offers in capturing CSO discharges. Because the tunnel capacity is available to all of the area served by the tunnel storage system, the system can be operated to optimize the water quality benefits associated with CSO control by prioritizing tunnel storage capacity for those basins where CSO control is most critical. The recommended CSO control plan would prioritize storage capacity for the Charles River Basin, an impounded river basin that is heavily used by sailors and rowers, and for Dorchester Bay, the area of Boston Harbor ringed by Boston's major salt water bathing beaches. Reserving tunnel storage capacity for these two basins means that there will be increased CSO discharges in Boston's Inner Harbor. Because activities in the Inner Harbor are more commercial and industrial, a slight decrease in CSO control in the Inner Harbor is perceived as a reasonable tradeoff for improved water quality in waters that are used more for recreational activities.

The recommended CSO control plan will reduce the volume of CSO discharges into Boston Harbor and its tributaries by 92% as compared with existing conditions. This reduction represents an 82% reduction when compared with the volume of discharge that will occur following the planned improvements to the Deer Island WWTF. The CSO control plan will also reduce the number of CSOs discharge events dramatically. The recommended system would reduce the number of overflows in each basin to no more than four per year. This compares with some 60 to 80 presently, and up to 70, even after the upgrades to the Deer Island WWTF. This reduction in volume will be accompanied by an equivalent reduction in pollutant loads.

13.5 POLLUTION CAUSED BY STORMWATER

While CSO discharges represent a major source of pollution, much of Boston Harbor and its tributaries would fail to meet WQS for bacterial pollution, even in the absence of CSOs. Other sources of pollution include urban runoff, illegal or illicit sanitary connections in storm sewers, contaminated sediments, upstream sources, improper or illegal use of marine toilets, and atmospheric deposition. The existence of other sources of pollution are a concern for the MWRA. Rate-payers who have funded an expensive CSO control program, which itself is only part of an even more expensive program to upgrade the sewerage system in the Boston area, will expect the result of that program to be

Table 5. Annual CSO and Stormwater Volumes[a]

Location	Existing Conditions	No Further CSO Action[b]	CSO Program	Stormwater Volume
Alewife/Mystic	228	71	1	3860
Charles River Basin[c]	4037	1975	356	8733
Boston Inner Harbor	6400	2512	402	4473
Dorchester Bay	859	438	116	332
Neponset River Estuary	39	20	0	2192

[a] All volumes in millions of gallons.
[b] "No Further CSO Action" refers to conditions once improvements to the Deer Island WWTF are completed but no further CSO control has been completed.
[c] Includes upper and lower Charles River Basin.

fishable and swimmable waters. However, unless action is taken to address other sources of pollution, swimming and fishing could still be limited in Boston Harbor following storms.

Estimates of the importance of each of these sources vary widely, and data on many of the sources are difficult to find. The MWRA's CSO facilities planning program did examine the importance of stormwater, and there are estimates on the importance of this source. Table 5 shows the volumes of stormwater in each water basin in the combined sewer area as compared with the volume of CSO discharged under existing conditions and if no CSO control, beyond that provided by the increased WWTF capacity, were provided. In all basins except Dorchester, the volume of stormwater vastly exceeds the volume of CSO that will be discharged following upgrades to the WWTF.

While the concentrations of pollutants found in stormwater are generally lower than that associated with combined sewage, the volumes of stormwater in most basins is large enough that modeling predicts significant pollution effects from the stormwater discharges. The estimated pollutant loads for Total Suspended Solids and BOD attributable to stormwater and CSOs are shown on Table 6. Modeling has shown that the bacterial pollutant loads would cause violations of the numeric bacterial standards even without CSO discharges. As shown on Table 7, in the Charles River, such a violation, attributable only to stormwater, would last for approximately 40 h following a 3-month storm. In the Alewife Brook, where stormwater vastly overwhelms CSO as a source of pollution, the violation of the bacterial standard would last for over 80 h.

13.6 U.S. EPA STORMWATER REGULATIONS

The U.S. EPA recently promulgated new stormwater regulations in res[to 1987 amendments to the U.S. Clean Water Act. The regulations estab

Table 6. Annual Pollutant Loads

Location	Total Suspended Solids[a]		
	Existing Conditions	No Further CSO Action[b]	Stormwater Loads
Alewife/Mystic	212	66	1835
Charles River Basin[c]	6494	3175	4151
Boston Inner Harbor	13163	5158	2126
Dorchester Bay	951	485	158
Neponset River Estuary	33	17	1042
	BOD[a]		
Alewife/Mystic	123	38	757
Charles River Basin[c]	2981	1459	1712
Boston Inner Harbor	5844	2290	877
Dorchester Bay	497	253	65
Neponset River Estuary	20	10	430

[a]   TSS and BOD in 1000 lb.
[b]   "No Further CSO Action" refers to conditions once improvements to the Deer Island WWTF are completed but no further CSO control has been completed.
[c]   Includes upper and lower Charles River Basin.

Table 7. Expected Duration of Bacterial Standard Violations Following a Three-Month Storm

Location	Hours of Standard Violations[a]		
	All Sources	CSOs Only	Stormwater Only
Alewife/Mystic	84	35	81
Charles River Basin[c]	72	66	41
Boston Inner Harbor	27	24	15
Dorchester Bay	37	36	21
Neponset River Estuary	43	12	43

[a]   Hours of violations of fecal coliform standard assuming that improvements to the Deer Island WWTF have been completed.

permitting program for industries and large and medium sized municipalities. Under the permit program municipalities will have responsibility for identifying, monitoring, and controlling pollutants from the storm sewer systems which serve virtually all developed communities. The requirements are imposed on medium and large municipalities; however, it seems reasonable to assume that smaller communities may ultimately face the same responsibilities.

The permit application process for municipalities has two parts calling for identification of stormwater outfalls: characterization of the discharges from those outfalls, and monitoring of discharges. In the Part 1 applications, municipalities are required to identify discharge locations and characterize the discharges. A large municipality must field screen up to 500 major outfalls or locations in a grid system describing the stormwater drainage or sample all major outfalls as part of the characterization of discharges. Grab samples must be collected when flow is observed during the screening. In the Part 2 application, the municipality must follow up its initial work with quantitative data for five to ten "representative" sampling points. The quantitative data must be gathered from at least three storm events and is intended to provide a basis for estimates of seasonal pollutant loads and of the event mean concentration.

Given EPA's focus on gathering of data and its commitment to high quality solutions, one might hope that these new regulations would encourage careful assessment of stormwater pollution and development of creative solutions. The regulations, however, may generate much more controversy than stormwater control. First, the cost of simply filing and maintaining permits under the regulations will be very high and may leave municipalities with little money or energy for implementing stormwater controls. Second, municipalities face potential legal liability for stormwater discharges that cause violations of WQS, and a municipality may be tempted to structure its monitoring to avoid finding the most serious of its stormwater discharges in order to minimize its risks.

The requirements for monitoring of stormwater discharges pose a serious problem for municipalities, particularly in these difficult financial times. Monitoring can be very helpful in understanding complex problems, and data generated through monitoring can ultimately help formulate creative solutions to difficult problems. However, the MWRA's experience in its CSO monitoring and sampling program demonstrates monitoring of wet-weather overflows is difficult and costly. The MWRA's sampling program had high mobilization costs due to many fruitless attempts to sample. While peculiar weather patterns were the principal cause of the sampling problems, experience with other sampling efforts since then has confirmed the difficulties of predicting the weather well enough to minimize mobilization costs while still obtaining necessary data during wet-weather discharges. Informal discussion of the stormwater regulations by U.S. EPA suggests that it envisions use of automatic sampling equipment as means of minimizing data collection costs. However, automatic samplers are limited in their usefulness, particularly in measuring bacterial pollutants. Even though the number of stormwater discharges to be monitored seems small, the costs of such monitoring may be very high.

The second problem with the regulations relates to the manner in which U.S. EPA will enforce the water quality requirements of the U.S. Clean Water Act. Even if municipalities find the funds to complete the required monitoring programs, as permittees, they face potentially enormous risks in submitting data. U.S. EPA Office of General Counsel has stated that, like permits for CSO discharges, permits for municipal and industrial separate storm sewer systems must include any requirements, in addition to the technology-based requirements, necessary to achieve compliance with WQS. The amendments to the U.S. CWA, which led to the development of the new permit program, specified that municipalities must control pollution from stormwater to the "maximum extent practicable" (MEP) (CWA Section 402(p)(3)(B)(iii)). This standard seems to be less than "best available technology" and "best conventional pollutant control technology" (BAT/BCT) which appear elsewhere in the U.S. CWA. However, control of stormwater to the "maximum extent practicable" may not be sufficient to meet WQS, and thus, a stormwater control program using that technology-based standard may not be consistent with permit requirements.

The risk that MEP control programs will not be sufficient to meet WQS is real. Data on pollutants found in urban runoff was gathered and compiled in a project, called the National Urban Runoff Program (NURP),[1] completed in the early 1980s. Findings from the NURP data indicate that concentrations of pollutants in urban runoff are highly variable, both from location to location and also from event to event. However, the data also show that coliform bacteria are present in levels high enough to cause violations of WQS criteria, even where there is substantial dilution of the runoff by the surface waters. The mean concentration of fecal coliform bacteria reported in the NURP Final Report was up to 50,000 GS/100 mL, well in excess of the Massachusetts WQS criteria of 200 organisms per 100 mL. The WQS violations shown by the MWRA's model, reported above, also predict substantial risk of violations.

The risk faced by a municipality due to failure to meet WQS is enhanced because resolution of priorities in pollution control programs is often resolved by confrontation, and litigation, among various interest groups, including municipalities. A municipality or project proponent cannot, or will not, put forward a project that will meet all requirements, perhaps because of cost, and public interest groups cannot, or will not, accept projects that fail to comply with the strict letter of the law, regardless of cost. During the 1970s and 1980s, there was no agreement in Massachusetts as to the appropriate goals for improvements to the antiquated and inadequate sewage treatment facilities. Consequently, no projects to reduce the sewage-related pollution in Boston Harbor were begun, despite several years of litigation. The deadlock was broken only when a new agency, the MWRA, was created, which dedicated itself to meeting the requirements of the law, whatever the cost. While the MWRA's approach has meant that decades of increasing pollution of Boston Harbor are finally over, it is reasonable to ask whether we as a society can afford to say that cost is no object in controlling pollution. Perhaps even more important, we must question whether attempts to force attainment of unreasonable high goals leads to a failure to make any progress at all. Certainly, that was a major factor in the long impasse on sewage treatment in Boston.

These questions raise serious issues for stormwater control. As with treatment of CSOs, treatment of stormwater to meet WQS presents enormous challenges. The very strict bacterial standards in the Massachusetts WQS will be difficult to meet if there is any appreciable level of bacterial pollution in the stormwater. The large volumes and high flow rates combined with the need to treat high levels of bacteria may require large and costly treatment facilities like those proposed for CSO control. Some have suggested that stormwater that is very polluted or that discharges into particularly sensitive water bodies should be sent to the MWRA's CSO storage system. That would result in significant pumping costs and even higher treatment costs for secondary treatment. Yet it is entirely possible that implementation of the management practices suggested by U.S. EPA in the stormwater regulations will lead to significant improvements in the quality of stormwater runoff and, hence, in the surface water quality. Nevertheless, even dramatic improvements in water quality may not be sufficient to meet WQS.

Water quality in Boston Harbor has improved considerably over the last few years, in large part due to improved maintenance and "housekeeping" practices by both the MWRA and the municipalities which operate the local sewer systems. If a municipality could target areas for structural and source control, it might find that relatively small expenditures bring about significant water quality improvements. In Boston, the MWRA and the Boston Water and Sewer Commission (BWSC), the agency responsible for storm and sanitary sewers in the city of Boston, found that targeting the identification and correction of illegal and illicit sanitary connections in selected storm sewers has yielded dramatically improved water quality conditions at some of Boston's heavily used beach areas. A strategy of identifying and correcting specific sources of pollution, rather than developing an areawide database for the stormwater discharges, is much more likely to bring about tangible improvements in surface water quality. Unfortunately, such a program would not comply with the new stormwater regulations. Instead, Boston may spend upwards of $1 million on data collection and permit applications, reducing the funds available for correcting identifiable problems in areas of critical concern. Moreover, if Boston does not demonstrate that the proposed controls will lead to compliance with WQS, it could find itself forced through litigation to adopt an expensive, albeit effective, stormwater treatment program, rather than investigating less expensive, but still reasonably effective management practices.

13.7 CONCLUSION

Much has been learned during CSO facilities planning about pollution from CSOs. The facilities proposed to control CSO discharges in Boston Harbor and its tributaries will provide a very high level of control for a very serious pollution problem. The plan is consistent with WQS which may be among the most strict in the country. Nevertheless, the implementation of a CSO control program represents only one of many steps in restoring the water quality of Boston Harbor. Stormwater and other sources of pollution must be identified and controlled. However, the MWRA's aggressive strategy for CSO control is not appropriate for control of stormwater. It is too expensive to be implemented for the enormous volumes of relatively less-polluted stormwater.

In searching for strategies to improve the environment, we as a society must balance the choices regarding environmental quality with our ability to both identify solutions and to pay for the implementation of those programs. We must also approach these debates from a spirit of cooperation and a willingness to consider new alternatives. For example, adoption of wet-weather water quality standards would go a long way towards eliminating the possibility that enforcement of strict numeric limits will lead to unreasonably expensive stormwater control program. And a willingness to move towards solutions step-by-step rather than in one giant leap will provide opportunities to assess the effectiveness of less expensive solutions and to reassess our goals based upon this greater understanding. With cooperation and knowledge, which is being fostered by this conference, we will be successful in meeting the goals of improved stormwater management.

REFERENCES

1. CH2M Hill Team. "Final Combined Sewer Overflow Facilities Plan and Final Environmental Impact," Report prepared for the Massachusetts Water Resources Authority, September 28, 1990.
2. "National Pollutant Discharge Elimination System Permit Application Regulations for Storm Water Discharges," *Federal Register*, Vol. 55, No. 222 (November 16, 1990).
3. Elliott, E.D. Internal U.S. EPA communication (January 9, 1991).
4. "Results of the Nationwide Urban Runoff Program," U.S. EPA Final Report (1983).

14   THE DETECTION AND DISINFECTION OF PATHOGENS IN STORM-GENERATED FLOWS

14.1 INTRODUCTION

Storm-generated flows occur on both an intermittent and random basis. During and after rainfall, these flows exhibit highly varying intensities over short periods of time with respect to both pollutant and microorganism quality and hydraulic quantity. In general, a sewer or channel can flow from completely dry to a thousand times the steady-state flow conditions associated with sanitary (domestic) wastewater. The characteristics of stormwater also vary according to the manner in which the stormwater is routed to the receiving water. Storm-generated discharges entering a body of receiving water can originate from separate storm sewers, from combined sewers carrying a mixture of sanitary wastewater and stormwater (combined wastewater), or from sanitary sewers inadvertently or illicitly cross-connected to separate storm sewers. In addition, receiving waters can contain discharges from both separate storm sewers and combined sewer systems from urban and/or nonurban land areas. In view of the many and varying factors that dictate the pollutant and microbial content of stormwater and/or their receiving waters, the adaptation of existing analytical and disinfection methods to evaluate and treat these microorganisms has proven difficult if not ineffective.[1]

For the control of microorganisms in storm flows two basic needs have arisen.[2] First, it is necessary to determine a storm flow's human pathogen content and pathogenicity along with the pathogens' relationships to certain indicators. In stormwater studies, total coliform (TC), fecal coliform (FC), and sometimes fecal streptococcus (FS) remain the traditional indicators of human pathogens. However, these indicators and their recommended limiting values have been adopted out of their routine use in potable water and sanitary wastewater analysis. Their appropriateness for the analysis of stormwater remains

0-87371-805-4/93/$0.00+$.50
© 1993 by Lewis Publishers, Inc.

questionable, particularly where stormwater does not enter the receiving water mixed with sanitary wastewater. For stormwater uncontaminated by sanitary wastewater, traditional fecal indicator levels may misrepresent the disease-causing potential of the stormwater, resulting in the premature closure of beaches and the unwarranted adoption of costly disinfection and control measures. In addition, a significant portion of swimming-related illnesses are associated with exposure to nonenteric pathogens, e.g., *Staphylococcus, Pseudomonas, Klebsiella,* and adenoviruses that can result in infections of the skin, ears, eyes, and upper respiratory tract, risks which cannot be estimated using fecal coliform densities alone.[3-6] In general, criteria based solely on TC or FC densities inadequately represent the actual human disease contraction potential, i.e., pathogenicity of a storm flow and its receiving water, causing a misguided concern over some disease hazards and the neglect of others. Epidemiological studies are severely lacking which specifically address the human pathogen potential of receiving waters fed by the stormwater runoff of various watersheds types.

Second, for stormwater discharges which pose serious health hazards, e.g., storm flows from combined sewers or from storm sewers containing a significant number of sanitary cross-connections or that drain watersheds containing feedlot operations, disinfection requirements and procedures should be designed to accommodate the unique characteristics of these flows. For example, stormwater's high volumes and flow rates require the development of high-rate disinfection systems to save on large tankage or dosage requirements, while the highly varying qualitative and quantitative character of these flows require flexible facility design and operational techniques in order to prevent unnecessary and costly disinfection expenditures.

14.2 BACTERIAL CRITERIA DEVELOPMENT — A HISTORICAL PERSPECTIVE

It has long been recognized that water can be a medium for pathogenic bacteria, fungi, and viruses and that the source of many of these disease-causing microorganisms is fecal contamination. Yet the difficulty and expense associated with the isolation and measurement of pathogenic microorganisms has resulted in the development of methods to monitor certain indicator organisms, i.e., microorganisms indicative of the presence of fecal contamination. Bacteria of the TC group became the generally accepted indicator for fecal pollution despite the fact that many of the bacteria in this group were known to be of a nonfecal origin.

Total coliform bacteria are gram-negative, nonspore-forming, and lactose-fermenting bacilli which produce gas within 48 h at 35°C.[7] These characteristics allow for delineation of the TC group which include many different genera, e.g., *Escherichia (E. coli), Citrobacter, Klebsiella,* and *Enterobacter.* Because the latter three genera are rarely associated with enteric wastes, attempts were made to narrow the scope of the TC assay to the detection of those organisms which originate solely from fecal wastes. The FC test, an elevated temperature (44.5°C) procedure used with modified media, was thus developed and became the preferred indicator assay for fecal pollution. The FC test selects primarily for *Klebsiella* and *E. coli* with infrequent positive results for other genera. However, *E. coli* is the only member of the FC group that is a consistent inhabitant of the intestinal tract of humans and other warm-blooded animals.[8] Thus, although the FC test is an improvement over the original TC test, it is still not specific to enteric bacteria in general and human-enteric bacteria in particular.

The most widely used bacteriological criterion in the U.S. today is the maximum recommended density of 200 FC colonies/100 mL of sample.[9] However, as a brief review of its adoption will illustrate, this criterion is not supported by either epidemiological or pathogenic-contact evidence.

Studies of gastrointestinal (GI) illness in swimmers in the early 1950's found that TC densities between 2300 and 2400 colonies/100 mL caused a significantly higher incidence of symptoms.[6,10] Later, as FC became the favored indicator for sanitary wastewater, early TC data collected on the Ohio River were reevaluated to determine a FC/TC ratio of approximately 0.18.[11] This ratio, plus a safety factor of 0.5, was applied to the TC densities (2300 to 2400 colonies/100 mL) known to produce health effects and an average criterion of 200 FC colonies/100 mL was generated.[11] This value was believed to provide bathers adequate protection from pathogenic contamination and was recommended by the U.S. Public Health Service in 1968.[9]

In 1973, a U.S. EPA publication cited studies by Geldreich, and Geldreich and Bordner which correlated the occurrence of the pathogen *Salmonella* with FC densities.[12-14] These studies found that the frequency of *Salmonella* detection increased sharply at FC densities above 200 colonies/100 mL, and reached a 97.6% detection maximum when FC densities exceeded 2000/100 mL.[13] On the basis of this and other data the EPA suggested a limit of 2000 FC colonies/100 ml for the protection of public (potable) water supplies but did not recommend a criterion for recreational waters due to the "paucity of valid epidemiological data".[12]

In a 1976 report, the EPA reinforced the original 1968 criteria of 200 FC colonies/100 mL for recreational waters despite numerous criticisms of its deficiencies.[9,15-19] The 1976 report acknowledged that epidemiological evidence to support the criterion was lacking but concluded that FC levels remained the best measure of microbiological water quality because of problems associated with the detection of other indicators or pathogenic microorganisms. Thus, despite the absence of epidemiological evidence, or an acceptable alternative indicator, TC and FC criteria were adopted and enforced throughout the country.

More recently, advances in microorganism isolation and identification have permitted researchers to study the relationship between swimming-associated illnesses and specific taxa of the FC group. In the early 1980s EPA conducted two such studies of both marine and freshwaters which aimed to determine the relationship between GI swimming-disorders and the bathing-water densities of FC, enterococci, and *E. coli*.[20,21] Each study used regression and correlation analysis to compare the strength of association of the various indicator bacteria to GI illness, thereby providing both an epidemiological rationale for the suggested criteria and the flexibility to consider other levels of risk.

On the basis of the correlation data the EPA marine study concluded that enterococci would be superior to *E. coli* as an indicator of fecal pollution at ocean beaches,[21] while the statistics generated in the freshwater study indicated that either enterococci or *E. coli* would be a suitable indicator for freshwater-bathing quality.[20] The results of these studies also revealed that due to differences in the die-off rate of indicator bacteria in freshwater and seawater, equivalent enterococci densities led to illness rates among swimmers in marine waters approximately three times greater than that observed amongst freshwater bathers. As stated in the freshwater study, this suggests criteria developed for freshwaters would be inappropriate if similarly applied to marine waters:[20]

> The significance of these findings is that a single water quality criterion for seawater and freshwater has been effectively eliminated from consideration, and therefore a separate criterion should be used for each type of bathing water.

Although the freshwater study found that both enterococci and *E. coli* densities displayed an excellent relationship to GI illness rates, *E. coli* exhibited the higher correlation coefficient and a lower standard error.[20] Additional factors favoring *E. coli* as the indicator of choice for freshwater bathing quality included: (1) its often higher density than enterococci both in human feces and sanitary-wastewater effluent,[22,23] and (2) its apparent hardiness in freshwater, relative to that of enterococci.[24]

The results of both studies clearly confirmed that the rate of GI illness increased with fecal contamination. However, in statistically evaluating the relationship between FC densities and GI disorders, both studies found that FC densities were unrelated to swimming-associated gastroenteritis.[20,21] Data from other studies were consistent with these findings.[25,26] The implication of these results was best summarized in the freshwater report:[20]

> Bacteria from sources other than the gastrointestinal tract of man and other warm-blooded animals, which fit the definition of fecal coliform ... are present at densities high enough to sufficiently eliminate the usefulness of fecal coliforms as an indicator of fecal contamination of surface waters.

A 1986 EPA publication on water quality criteria addressed the limitations associated with the use of TC and FC indicators in the measurement of bathing water quality and recommended that states "begin the transition process to the new (*E. coli* and enterococci) indicators".[27]

The preceding chronology provides the background and rationale for current regulations regarding bacteriological water quality. In general, descriptions of adverse public health impacts resulting from the discharge of sanitary wastewater without prior treatment have gradually evolved from simple mathematical correlations to EPA's current risk assessment approach. However, despite the lack of correlation between TC and FC levels and swimming-related illnesses and the 1986 EPA recommendation for the adoption of new recreational water quality criteria, many states still retain the TC and FC criteria first recommended in 1968.

In the search for a more accurate determination of the nature of the pollution source and thus a measure of the human disease potential of the receiving waters, several indicator relationships and/or microbial detection methods for pathogenic bacteria or human enteric viruses have been examined: FC/FS ratios;[28] *P. aeruginosa*/FC ratios;[29] *Clostridium perfringens* and its relation to FC densities;[26,30] fecal sterols (e.g., epicoprostanol and coprostanol);[31,32] species-specific bacteriophages (e.g., RNA coliphages[33], *Bacteroides fragilis* phages,[34,35] etc.); and some species of the genus Bifidobacteria.[30,36] Investigation and evaluation of several of these alternative indicators have shown them to either fall short of the list of requirements commonly cited for indicators,[37,38] or to possess only limited usefulness.

Recent methods allowing the direct detection of waterborne pathogens include gene probes and Polymerase Chain Reaction (PCR) techniques. PCR and DNA probe methods have already been used in the rapid detection and enumeration of coliform bacteria, *E. coli*, *Shigella* spp.;[39-41] *Salmonella* spp.;[42] and *Giardia*;[43] and it is anticipated that these methods will eventually be applied to the direct detection of human enteric viruses.[44]

14.3 STORMWATER QUALITY AND ITS RELATIONSHIP TO HUMAN DISEASE POTENTIAL

Despite ongoing research on alternative indicators, the common bacterial indicators for recreational waters remain TC and FC. For receiving waters contaminated by *sanitary* wastewater alone or in combination with stormwater the choice of either FC or TC densities as an indicator of pathogens may be a satisfactory one.

For separate storm drainage systems and streams that are not separated from the sources of human fecal contamination, e.g., sanitary wastewater, the results of microbiological analyses suggest that these waters can and do present a potential health hazard. Some of the disease-causing microorganisms isolated from stormwater runoff and urban streams include enteroviruses (e.g., poliovirus, Coxsackie B virus, and Echovirus) and bacteria in the form of *Pseudomonas aeruginosa*, *Staphyloccus aureus*, and *Salmonella*.[37,45,47] For example, Table 1 lists pathogen and indicator densities found in separate storm sewers (containing varying extents of sanitary wastewater inflows) in the Baltimore, Maryland area, while Table 2 shows the microbial results of sheetflows collected from the Emery and Thistledown catchments in the Toronto, Canada area.

TABLE 1. Geometric Mean Densities of Selected Pathogens and Indicator Microorganisms in Stormwater[37]

Sampling Station	Enterovirus PFU/10 L	Salmon. sp. MPN/10 L	P. aeruginosa MPN/10 L	Staph. aureus MPN/100 mL	TC MPN/100 mL ($\times 10^4$)	FC MPN/100 mL ($\times 10^3$)	FS No./100 mL ($\times 10^4$)	Enterococci No./100 mL ($\times 10^4$)
Stoney Run[a]	190	30	1300	12	4.8	19	4.1	1.4
Glen Ave.[b]	75	24	3300	14	24	81	66	21
Howard Park[c]	280	140	5200	36	120	450	24	5.9
Jones Falls[b]	30	25	6600	40	29	120	28	8.7
Bush St.[d]	6.9	30	2000	120	38	83	56	12
Northwood[d]	170	5.7	590	12	3.8	6.9	5	2.1

[a]   Three sanitary bleeders (intentional sanitary sewage overflows from interceptors).
[b]   One sanitary bleeder.
[c]   Combined sewer.
[d]   Storm only.

TABLE 2. Sheetflow Quality Summary Emery and Thistledown
Catchments, Toronto, Canada[47]

Sheetflow Area	Median and (Range) Densities (1000 bacteria/100 mL)		
	FC	FS	P. aeruginosa
Unpaved driveways, storage areas			
Emery	26	6.2	0.5
	(0.02–300)	(0.18–22)	(0.02–51)
Roof runoff			
Emery	1.6	0.69	0.05
	(0.56–2.6)	(0.38–1.0)	(<0.02–0.1)
Thistledown	0.5	0.94	0.1
	(0.12–3.7)	(0.54–5.1)	(0.02–90)
Sidewalks			
Emery	55	3.6	3.6
	(19–90)	(3.3–3.9)	(0.1–7.1)
Thistledown	11	1.8	0.6
Paved parking/ storage and driveways			
Emery	2.8	0.9	0.7
	(0.03–66)	(<0.1–39)	(0.02–15)
Thistledown	2.0	1.5	0.52
	(0.1–980)	(<0.1–690)	(0.08–5)
Paved roads			
Emery	19	8.5	5.4
	(1.8–430)	(0.6–240)	(1–15)
Thistledown	4.8	7.9	0.1
	(0.8–15)	(1.1–13)	(0.02–1.7)
Overall Range	0.02–980	<0.1–690	<0.02–90

In stormwater flows where pathogen concentrations were significant
and could not be correlated to storm events or soil populations, the most
frequently cited sources of the contamination were sanitary wastewater line
leaks, interceptor diversions, or intentional cross-connections into the storm
drainage system,[37,45,46] i.e., a lack of total separation from sanitary wastewater
sources. The Baltimore, Maryland study determined that the frequency of
pathogenic contamination could be directly related to the extent of sanitary-
wastewater diversions or number of direct connections into the stormwater
system.[37] The analyses of dry-weather base flows in separate stormwater

drainage systems can often determine the extent of contamination by sanitary wastewater via illicit or inadvertent cross connections.[48] As an example, a survey conducted in Toronto, Canada found that dry-weather-base flows in the separate stormwater drainage system exhibited statistically similar FC populations to those observed in stormwater runoff,[47] implying the presence of a continuous microbial pollutant source. Despite evidence of pathogenic contamination of stormwater, it has been argued that the presence of these pathogens in stormwater does not, in fact, constitute a significant health hazard.[37] This argument cites the low densities of pathogenic microorganisms observed in urban storm runoff, the further dilution of these flows upon reaching recreational waters, and the large infective doses of bacteria such as selected species of Salmonella (10^5 organisms) in concluding that any threat to swimmers should be small, "since prodigious swallowing of water would be required in order to increase the risk of enteric disease".[37] Unfortunately, the evidence of low densities coupled with high infective doses cannot minimize the health hazard of pathogens, such as *P. aeruginosa*, *Salmonella typhosa*, *Shigella*, or enteroviruses, that either do not require ingestion for infection or require very low infective doses. However, due in part to past difficulties in the isolation and quantification of some of these species, particularly at the low densities normally observed in storm- and receiving waters, there has been little study of their correlation with swimming-associated illnesses.

For example, several studies have found large (10^3 to 10^4 organisms/100 mL) populations of *P. aeruginosa* in urban streams and stormwater runoff.[37,45,47] PA/FC ratios ranged over three orders of magnitudes (from 0.01 to more than 20), indicating that FC populations were poorly related to the density of this pathogen. The predominance of *P. aeruginosa* in stormwater coupled with its association with diseases transmitted through water contact, e.g., skin and ear infections, signifies its potential importance in evaluating the health hazard of waters receiving storm runoff. However, studies which have attempted to correlate PA densities to illness rates have reported only its poor relation to acute GI distress or total illness rates.[3,4,21] Little information is currently available regarding its correlation to body contact illnesses due to stormwater exposure, despite the suggested greater risk associated with this mode of transmission.

In 1977, it was estimated that 14.4% of urban areas containing 25.2% of the urban population was served by combined sewers.[49] These percentages have since declined due to the ongoing development of suburban communities that are either served by separate storm sewers or are unsewered, and the lessening of combined sewer construction. It has been well established that the bacteria isolated in stormwater runoff are predominately from nonhuman sources.[45,50] Thus, for receiving waters accepting separate stormwater inflows, a reliance on TC or FC indicators to determine bathing water quality may prove ineffective due the inability of this method to distinguish human from nonhuman, and possibly nondisease causing, sources (e.g., vegetation, soil, and animals).[2,3,28,37,51,52]

Several studies have isolated animal-associated enteric viruses and bacteria that can be transmitted to humans, e.g., *Yersinia, Cryptosporidium,* and *Salmonella,* in stormwater or surface waters in urban, rural, and agricultural watersheds, indicating that the disease-causing potential of these sources cannot be neglected.[50,53-55] However, to date few epidemiological studies have attempted to correlate incidences of GI or total illness with FC densities arising primarily from nonhuman sources, e.g., stormwater runoff uncontaminated by sanitary wastewater. Such studies, undertaken for a variety of watershed types, are necessary to insure that the continued reliance upon coliform indicators to determine water quality criteria for stormwater receiving-recreational-waters does not erroneously hinder their recreational usage.

To date only one well-documented study has been conducted that has addressed diseases that may result from direct contact with bathing waters whose sole source was rainwater runoff from a (forested) watershed.[56] This study used epidemiological data to compare the health status of swimmers utilizing the waters during wet-weather periods with that of nonswimmers. The study site was located in a semirural community and consisted of a 3-acre freshwater pond with no known source of human fecal contamination. During a 49-day period water samples were collected three times daily and analyzed for *E. coli,* enterococci, *P. aeruginosa,* staphylococci, and FC. Data on rainfall, bather density, and the occurrence of GI illness among the monitored families were also collected.

Monitoring results indicated that the geometric mean densities of *E. coli* and FC were over two times greater on rain days than on nonrain days, while for enterococci the density ratio for rain/nonrain days was four. These three fecal-related indicators also exhibited significant correlation with each other, i.e., when one increased in density, the other two also increased. No correlation was observed between indicator bacteria levels and bather density. Conversely, staphylococci densities were related to bather density but not to any of the fecal indicator bacteria or to rainfall. Health data were analyzed by pairing each swimmer illness with the indicator density associated with the day of exposure and then segregating these illnesses into two groups based on high and low parameter densitites (Table 3). GI illness was observed to be strongly associated with swimming, but illnesses appeared randomly dispersed in the high and low indicator groups, suggesting that no association exists between GI illness and high fecal indicator bacteria densities. However, a significant association was observed between swimming-associated illnesses and high densities of staphylococci or high densities of bathers. The authors concluded that the reported illnesses were probably due to agents transmitted from swimmer to swimmer, and was not related to pollution discharged into the pond during wet weather. The high densities of the three fecal indicators, which could be correlated with daily rainfall levels, were attributed to the presence of warm-blooded animals in the wooded areas surrounding the swimming pond.[56]

The results of this study are consistent with an earlier work which documented GI and total illness among 8400 swimmers and nonswimmers at ten freshwater

TABLE 3. Association Between Cases of GI Illness and Various
Monitored Parameters[56]

Monitored Relative Parameters	High Groups		Low Groups		
	High Values	Illnesses[a]	Low Values	Illnesses[a]	Risk
Rainfall[b]	≥ 0.2	29	<0.2	14	2.1
Enterococci[c]	≥ 20	35	<20	18	1.9
E. coli	≥ 75	29	<75	21	1.4
FC[c]	≥ 80	34	<80	20	1.7
Bathers	≥ 53	29	<53	6	4.8
Staphylocci[c]	≥ 45	31	<45	12	2.6

[a]   Number of illnesses per 1000 person-days.
[b]   Inches per day.
[c]   Density per 100 mL.

beaches in Ontario using total staphylococci, fecal streptococci, fecal coliform, heterotrophic bacteria, and *P. aeruginosa* as indicators.[3,4] The findings indicated total staphylococci densities possessed the strongest dose-response relationship and proved to be the most consistent indicator of total illness as well as eye and skin disease.

14.4 DISINFECTION

Disinfection requirements and associated facilities in the U.S. were established and designed to protect waters used for recreational, shellfishing, and potable supply purposes by controlling TC or FC bacteria at various levels. For the treatment of municipal wastewater, the regulations are often expressed as a combination of technology-based definitions (e.g., dosage-contact time) and water quality-based criteria (e.g., coliform density) and may be applied on a continuous or seasonal basis. However, storm-generated flows containing human-fecal contamination require a different disinfection approach since these flows are intermittent, often high rate, and normally display wide seasonal and intra-/interstorm variations in quantity, temperature, and pollutant and bacterial characteristics. These unique characteristics of storm-generated flows necessitate the adoption of cost-effective, high-rate disinfection practices and the use of disinfection facilities that can be adaptable to both intermittent use and varying dosage requirements.[57,58]

Several factors determine the overall effectiveness of disinfection by chemical or other means. These include, but are not limited to (1) the nature and concentration of both the disinfectant and the products formed in the water after reaction with it, (2) the condition of the water, e.g., its suspended-solids and chemical characteristics, temperature, and pH, (3) the contact time between the disinfecting agent and the pathogen, (4) the mixing intensity imparted to the water, and (5) the nature and density of pathogens and their resistance to inactivation by the disinfectant used.[59] In general, lowering the pH or increasing the disinfectant concentration, the water temperature, the mixing intensity, or the contact time will increase a chemical disinfectant's effectiveness. High-rate disinfection, i.e., decreased disinfectant contact time, can be achieved through the use of one or more of these practices alone or in combination with each other.[57]

14.4.1 Chemical Disinfection

Due to their low cost, ease of use, and germicidal properties, elemental chlorine and chlorine compounds have traditionally been the most widely used chemical disinfectants. Early disinfection practices using these materials were generally confined to the treatment of sanitary wastewater and potable water after it passed through a treatment facility. These plants normally employed relatively large chlorine-contact tanks and achieved bacterial and viral kill by the addition of elemental chlorine (Cl_2), or sodium hypochlorite ($NaOCl$).[60] Stormwater and combined sewer overflows (CSO) were generally not treated in a comparable manner due to the assumption that disinfection of dry-weather flow provided effective protection to the receiving waters. This assumption was challenged in the 1960s by several studies which specifically addressed the treatment and disinfection of stormwater and CSO, and which recognized that large variations in flows were the principal problems to be overcome in its effective chlorination.[61-63]

Throughout the 1970s, EPA research efforts were underway to determine the unique disinfection needs of stormwater and CSO.[64-68] Several of these studies employed the screening technique of Microstraining[R] to remove and/or fragment particulate and organic matter containing bacteria.[65-68] By reducing particulate size, the number of bacteria and viruses occluded within larger particulates, and consequently shielded from chemical attack, could be minimized. Coliform reductions across the microstrainer were found to be minor; however, it was confirmed that microstrained effluent was more amenable to disinfection, exhibiting a lower Cl_2 demand and requiring shorter detention times. These studies and others also addressed the importance of maximizing mixing intensity within the disinfection chamber to insure dispersion of the added disinfectant and increase the number of collisions between the bacteria and disinfectant.[69-74] Increased mixing intensity was shown to be achievable either statically by baffles, corrugated narrow pathways, or helical vanes; or dynamically by moving impellers. In general, the utilization of such high-rate mixing techniques within the chlorine contact chamber insured plug flow conditions, full residence times, and high liquid velocity gradients.

14.4.2 Alternative Disinfection Techniques

Since the 1970s, the growing awareness of the adverse environmental impacts associated with the chemical byproducts of continuous chlorination has led to increasingly restrictive residual chlorine requirements and has resulted in the employment of reducing agents or other dechlorinationtechniques at many chlorine treatment facilities.[75-78] Efforts to minimize these environmental risks and reduce the increasing chemical demands and contact times that are necessitated by the dechlorination procedure have fostered a strong interest in alternative disinfection technologies.

One disinfection technique which promises short detention times and the absence of toxic by-products is that of disinfection by ultraviolet (UV) light irradiation.[79-83] This technique works on the principle that all microorganisms that contain nucleic acids are susceptible to damage through the absorption of radiation in the UV energy range. However, the exact extent of damage, mutation, or death will depend upon an organism's resistance to radiation penetration, which will depend upon several factors including cell wall composition and thickness.[59]

UV-dosage requirements for achievement of target indicator concentrations depend upon several parameters including the frequency and intensity of the UV radiation, the number and configuration of the UV lamps, the distance between the wastewater and the lamp surface, the chamber turbulence, and a wastewater's absorption coefficient and exposure times.

Early studies of the UV disinfection process utilizing the maximum UV exposure levels then available indicated the limitations of the process:[79] disinfection of effluents containing high solids and dissolved organics proved largely ineffective. The high absorption of UV radiation by these substances served to attenuate the available UV energy, resulting in a reduction of its depth of penetration into the wastewater.[79] Consequently, organisms contained within large particulates experienced little or no irradiation due to the complete absorption of the radiation by the outer, protective layer. UV radiation doses have since been increased through improvements in system design and available equipment. However, coliform reductions have still remained highly dependant upon water quality. The high concentrations of solids and/or organics in primary effluent currently restricts the practical use of disinfection by UV irradiation to the treatment of secondary or tertiary effluent.[82-85] However, this disinfection technique, with its absence of toxic residuals and contact times on the order of seconds rather than minutes, would be a desirable choice for the treatment of the high flow rates associated with stormwater and CSO. New York City is currently evaluating a large-scale pilot plant proposal that would investigate the primary treatment efficiencies of CSO by improved UV irradiation techniques. The newly developed system would employ higher pressure lamps that emit higher intensity radiation and a broader spectrum of UV wavelengths.[86] Laboratory evaluation of a higher pressure lamp has demonstrated that its germicidal effectiveness is approximately ten times that of a conventional low pressure lamp,[87] effectively offsetting the ten fold increase in the number of lamps that would normally be required for the disinfection of CSO at efficiencies currently being met for treated effluents. Still more recent experiments have compared the germicidal effectiveness of modulated UV light with that of non-modulated UV radiation of the same intensity and exposure times. After exposure to pulsed radiation, viable bacterial populations were reported to number approximately 100-fold less than the populations observed after similar exposure to UV light that lacked modulation.[88]

14.6 CONCLUSIONS AND RECOMMENDATIONS

EPA's current emphasis on water-use attainability and risk assessment warrants the reevaluation of existing disinfection requirements and bacteriological criteria. Recent information relating incidence of disease in recreational water users to instream densities of various indicator microorganisms indicate that current criteria may be inappropriate at correctly assessing the risk to the public health. This is especially true for those receiving waters complicated by the influence of storm-induced inflows. Water quality and disinfection criteria for pathogenic bacteria and viruses are clearly needed for stormwater and CSO. To develop the former, regulatory agencies must address the problems of choosing the most appropriate indicator(s) and establishing acceptable limiting levels. In order to be most effective, microbial criteria should be derived from direct pathogen and epidemiological analyses for relating risk to a given level of protection. Where sources other than sanitary wastewater flows, e.g., stormwater runoff, enter bathing waters, the current criteria expressed in terms of FC organisms do not have such a basis.

The results of past epidemiological studies strongly suggest that current coliform-based indicator systems cannot be used to accurately assess the pathogenicity of recreational waters receiving stormwater from uncontaminated separate storm sewers or surface water runoff. Since the predominant bather-associated risk has been reported to be infections of the skin, ear, eye, and/or upper respiratory system, epidemiological guidelines are also required that address the presence of nonenteric pathogens. The adoption of multiple indicators (e.g., enteric and nonenteric bacteria) or alternative fecal indicators whose densities can be correlated with nonenteric infections may be necessary to provide a more accurate estimate of the total health risk associated with stormwater contact. In general, further epidemiological data, from a variety of watershed areas, is needed in order to determine which bacteria or viruses exhibit the best correlation with total illness rates. As was shown in the EPA's early freshwater and marine water studies, it cannot be expected that one indicator or indicator system will prove useable in all watershed areas and under all conditions.

For storm sewers which contain evidence of human fecal contamination, and thereby indicate the presence of illicit cross-connections, a serious effort should be directed towards identifying and eliminating these sanitary-wastewater sources.[32,33] However, should the long-range goal of a cross-connection elimination project include the expansion of a water body's recreational usage, the adoption of new water quality criteria based on epidemiological data should be simultaneously investigated. In general, if no consideration is given to the normally high coliform bacteria densities associated with stormwater, then costly programs aimed at eliminating human fecal contamination or its sources may prove ineffective in achieving the ultimate goal of swimmable waters.

In stormwater outfalls where cross-connections are too numerous or too costly to be corrected, it may be advisable to deal with the separate storm sewer as, in fact, a combined sewer.[80] Where stormwater is transported via combined sewers, the number of raw-sanitary-wastewater overflows into the urban waterways can be minimized through sewer maintainence, increased sewer carrying capacities, and the temporary containment of overflows for later (dry-weather) treatment.[38]

REFERENCES

1. Field, R., and E.J. Struzeski, Jr. "Management and control of combined sewer overflows," *J. Water Pollut. Control Fed.* 44(7):1393 (1972).
2. Field, R. "Microorganisms in Urban Stormwater — A U.S. Environmental Protection Agency Program Overview, Proceedings of the Workshop on Microorganisms in Urban Stormwater," U.S. EPA Report EPA-600-2-76-244, NTIS No. PB-263-030 (November 1976).
3. Seyfried, P.L., R.S. Tobin, N.E. Brown and P.F. Ness. "A prospective study of swimming-related illness. I. Swimming-associated health risk," *Am. J. Public Health* 75(9):1068 (1985).
4. Seyfried, P.L., R.S. Tobin, N.E. Brown and P.F. Ness. "A prospective study of swimming-related illness: morbitidy and the microbiological quality of water," *Am. J. Public Health* 75(9):1071 (1985).
5. Favero, M.S. "Microbiological indicators of health risks associated with swimming," *Am. J. Public Health* 75(9):1051 (1985).
6. Stevenson, A.H. "Studies of bathing water quality and health," *J. Am. Public Health Assoc.* 43:529 (1953).
7. Clesceri, L.S., A.E. Greenberg and R.R. Trussell, Eds. "Standard Methods for the Examination of Water and Wastewater," 17th ed. (Washington, D.C.: American Public Health Association, 1989).

8. Dufour, A.P. "Escherichia coli: the fecal coliform," in *Bacterial Indicators/Health Hazards Associated with Water*, A.W. Hoadley and B.J. Butka, Eds. (Philadelphia: ASTM, 1977), p. 48.
9. "Water Quality Criteria," National Tech. Advisory Committee Report to the Federal Water Pollution Control Administration, U.S. Department of Interior, Washington, D.C. (1968).
10. Smith, R.S., T.D. Woolsey and A.H. Stevenson. "Bathing Water Quality and Health. I. Great Lakes," U.S. Public Health Service, Cincinnati, OH (1951).
11. Geldreich, E.E. "Sanitary Significance of Fecal Coliforms in the Environment," Water Pollution Control Research Series No. WP-20-3, U.S. Government Printing Office 1966.
12. "Water Quality Criteria,1972," U.S. EPA Report-EPA-R3-73-003, Washington, D.C. (1973).
13. Geldreich, E.E. "Applying bacteriological parameters to recreational water quality," *J. Am. Water Works Assoc.* 62:113 (1970).
14. Geldreich, E.E., and R.H. Bordner. "Fecal contamination of fruits and vegetables during cultivation and processing for market — A Review," *J. Milk Food Technol.* 34(4):184 (1971).
15. "Quality Criteria for Water," U.S. EPA Report-EPA-440/9-76-023, NTIS No. PB-83-25994 (1976).
16. Henderson, J.M. "Enteric disease criteria for recreational waters," *J. San. Eng. Div.* 94:1253 (1986).
17. Moore, B. "The case against microbial standards for bathing beaches," in *International Symposium on Discharge of Sewage from Sea Outfalls*, (London: Pergamon Press, 1975), p. 103.
18. Cabelli, V.J., M.A. Levin, A.P. Dufour and L.J. McCabe. "The development of criteria for recreational waters," in *International Symposium on Discharge of Sewage from Sea Outfalls* (London: Pergamon Press, 1975), pp. 63–73.
19. Foster, D.H., N.B. Hanes and S. M. Lord, Jr. "A critical examination of bathing beach standards," *J. Water Pollut. Control Fed.*, 41:2229 (1971).
20. Dufour, A.P. "Health Effects Criteria for Fresh Recreational Waters," U.S. EPA Report-EPA-600/1-84-004 (1984).
21. Cabelli, V.J. "Health Effects Criteria for Marine Recreational Waters," U.S. EPA Report-600/1-80-031 (1983).
22. Slanetz, L. "Numbers of enterococci in water, sewage and feces determined by the membrane filter technique with improved medium," *J. Bacteriol.* 74:591 (1957).
23. Miescier, J., and V.J. Cabelli. "Enterococci and other microbial indicators in municipal wastewater effluents," *J. Water Pollut. Control Fed.* 54:1599 (1982).

24. Hanes, N., and R. Fragloa. "Effect of seawater concentration on the survival of indicator bacteria," *J. Water Pollut. Control Fed.* 39:97 (1967).
25. Dutka, B.J. "Coliforms are an inadequate index for water quality," *J. Environ. Health* 36:39 (1973).
26. Fujioka, R.S., and L.K. Shizumura. "Clostridium perfringens, a reliable indicator of stream water quality," *J. Water Pollut. Control Fed.* 57:986 (1985).
27. Dufour, A., and R. Ballentine. "Ambient Water Quality Criteria for Bacteria," U.S. EPA Report-EPA-440/5-84-002 (1986).
28. Geldreich, E.E., and B.A. Kenner. "Concepts of fecal streptococci in stream pollution," *J. Water Pollut. Control Fed.* 41(8):R336 (1969).
29. Cabelli, V.J., H. Kennedy and M.A. Levin. "Pseudomonas aeruginosa-fecal coliform relationships in estuarine and fresh recreational waters," *J. Water Pollut. Control Fed.* 48(2):367 (1976).
30. Cabelli, V.J. "Obligate anaerobic bacterial indicators," in *Indicators of Viruses in Water and Food*, (Ann Arbor, MI: Ann Arbor Science, 1978), p. 171.
31. Eganhouse, R.P. "Use of molecular markers for the detection of municipal sewage sludge at sea," *Mar. Environ. Res.* 25(1):1 (1988).
32. Walker, R.W., C.K. Wun and W. Litsky. "Coprostanol as an indicator of fecal pollution," *CRC Crit. Rev. Environ. Control* 12(2):19 (1982).
33. Kott, Y., N. Roze, S. Sperber and N. Betzer. "Bacteriophages as viral pollution indicators," *Water Res.* 8:165 (1984).
34. Jofre, J., A. Bosch, F. Lucena, R. Girones and C. Tartera. "Evaluation of Bacteroides fragilis bacteriophages as indicators of the virological quality of water," *Water Sci. Technol.* 18:167 (1986).
35. Tartera, C., F. Lucena and J. Jofre. "Human origin of Bacteroides fragilis bacteriophages present in the environment," *Appl. Environ. Microbiol.* 55(10):2696 (1989).
36. Resnick, I.G., and M.A. Levin. "Assessment of bifidobacteria as indicators of human fecal pollution," *Appl. Environ. Microbiol.* 42(3):433 (1981).
37. Olivieri, V.P., C.W. Kruse, K. Kawata and J.E. Smith. "Microorganisms in Urban Stormwater," U.S. EPA Report-EPA-600/2-77-087, NTIS No. PB-272245 (1977).
38. Vivian, C.M.G. "Tracers of sewage sludge in the marine environment: a review," *Sci. Total Environ.* 53:5 (1986).
39. Bej, A.K., R.J. Steffan, J. DiCesare, L. Haff and R.M. Atlas. "Detection of coliform bacteria in water by polymerase chain reaction and gene probes," *Appl. Environ. Microbiol.* 56(2):307 (1990).
40. Bej, A.K., J. DiCesare, L. Haff and R.M. Atlas. "Detection of Escherichia coli and Shigella spp. in water by using the polymerase chain reaction and gene probes for *uid*," *Appl. Environ. Microbiol.* 57:1013 (1991).

41. Bej, A.K., S.C. McCarty and R.M. Atlas. "Detection of coliform bacteria and Escherichia coli by multiplex polymerase chain reaction: comparison with defined substrate and plating methods for water quality monitoring," *Appl. Environ. Microbiol.* 57(8):2429 (1991).

42. Knight, I.T., S. Shults, C.W. Kaspar and R.R. Colwell. "Direct detection of salmonella spp. in estuaries by using a DNA probe," *Appl. Environ. Microbiol.* 56(4):1059 (1990).

43. Abbaszadegan, M., C.P. Gerba and J.B. Rose. "Detection of giardia cysts with a cDNA probe and applications to water samples," *Appl. Environ. Microbiol.* 57:927 (1991).

44. Francis, J., and L. Ellis. "Alternative indicators of human fecal contamination of stormwater, ongoing urban waste management and research center," University of New Orleans Project No. 92-A-010.

45. Qureshi, A.A., and B.J. Dutka. "Microbiological studies on the quality of urban stormwater runoff in southern Ontario," Canada, *Water Res.* 13:977 (1979).

46. Davis, E.M. "Maximum Utilization of Water Resources in a Planned Community — Bacterial Characteristics of Stormwater in Developing Rural Areas," U.S. EPA Report-EPA-600/2-79-050f, NTIS No. PB 80-129 091 (1979).

47. Pitt, R., and J. McLean. "Toronto area watershed management strategy study; Humber River pilot watershed project, Final report," The Ontario Ministry of the Environment, Toronto, Ontario (1986).

48. Pitt, R. Lalor, M., M. O'Shea and R. Field, "U.S. EPA's manual of practice for the investigation and control of cross-connection pollution into storm-drainage systems," Proceedings of the International Conference on Integrated Stormwater Management, Singapore (July 11–13, 1991).

49. Sullivan, R.H., M.J. Manning, J.P. Heaney, W.C. Huber, M.A. Medina, Jr., S.J. Nix and S.M. Hasan. "Nationwide Evaluation of Combined Sewer Overflows and Urban Stormwater Discharges, Volume I: Executive Summary," U.S. EPA Report-EPA-600/2-77-064a, NTIS No. PB 273-133 (Sept. 1977).

50. Geldreich, E.E., L.C. Best, B.A. Kenner and Van D.J. Donsel. "The bacteriological aspects of stormwater pollution," *J. Water Pollut. Control Fed.* 40(11):1861 (1968).

51. Field, R., and R., Turkeltaub. "Urban runoff receiving water impacts: program overview," *J. Environ. Eng. Div. ASCE* 107(EE1):83 (1981).

52. Rivera, S.C., T.C. Hazen and G.A. Toranzos. "Isolation of fecal coliforms from pristine sites in a tropical rain forest," *Appl. Environ. Microbiol.* 54(2):513 (1988).

53. Geldreich, E.E. "Bacterial populations and indicator concepts in feces, sewage, stormwater and solid wastes," in *Indicators of Viruses in Water and Food*, G. Berg, Ed. (Ann Arbor, MI: Ann Arbor Science, 1978), p. 51.

54. Fukushima, H. and M. Gomyoda. "Intestinal carriage of Yersinia pseudotuberculosis by wild birds and mammals in Japan," *Appl. Environ. Microbiol.* 57(4):1152 (1991).

55. Madore, M.S., J.B. Rose, C.P. Gerba, M.J. Arrowood and C.R. Sterling. "Occurance of Cryptosporidium oocysts in sewage effluents and selected surface waters," *J. Parisitol.* 73:702 (1987).

56. Calderon, R.L., E.W. Mood and A.P. Dufour. "Health effects of swimmers and nonpoint sources of contaminated water," *Int. J. Environ. Health* 1:21–31 (1991).

57. Weisman, D.A., and R. Field. "A Planning and Design Guidebook for Combined Sewer Overflow Control and Treatment," U.S. EPA Report-EPA-600/2-82-08 (September 1981).

58. Field, R. "State-of-the-art update on combined sewer overflow control," *CRC Crit. Rev. Environ. Control* 16(2):147 (1986).

59. Weber, W., Jr. *Physiochemical Processes for Water Quality Control,* (New York: John Wiley and Sons, 1972).

60. Steffensen, S.W., and N. Nash. "Hypochlorination of wastewater effluents in New York City," *J. Water Poll. Control Fed.* 39(8):1381 (1967).

61. Dunbar, D.D., and J.G.F. Henry. "Pollution control measures for stormwaters and combined sewer overflows," *J. Water Pollut. Control Fed.* 38(1):9 (1966).

62. Camp, T.R. "Chlorination of mixed sewage and storm water," *J. San. Eng. Div., Proc. ASCE* 87:1 (1961).

63. Evans, F.L., E.E. Geldreich, S.R. Weibeland and G.G. Robeck. "Treatment of urban stormwater runoff," *J. Water Pollut. Control Fed.* 40(5):R162 (1968).

64. Dow Chemical Co. "Chemical Treatment of Combined Sewer Overflows," U.S. EPA Report-11023FDB09/70, NTIS No. PB-199-070 (September 1970).

65. Cochrane Division, Crane Co. "Microstraining and Disinfection of Combined Sewer Overflows," U.S. EPA Report - 11023EVO06/70, NTIS No. PB-195-674/BA (June 1970).

66. Glover, G.E., and G.R. Herbert. "Microstraining and Disinfection of Combined Sewer Overflows-Phase II," U.S. EPA Report-EPA-R2-73-124, NTIS No. PB-219-879 (January 1973).

67. Maher, M.B. "Microstraining and Disinfection of Combined Sewer Overflows — Phase III," U.S. EPA Report-EPA-670/2-74-049, NTIS No. PB-235-771 (August 1974).

68. Diaper, E.W.J. and G.E. Glover. "Microstraining of combined sewer overflows," *J. Water Pollut. Control Fed.* 43(10):2101 (1971).

69. Pontius, U.R., E.H. Pavia and D.G. Crowder. "Hypochlorination of Polluted Stormwater Pumpage at New Orleans," U.S. EPA Report-EPA-670/2-73-067, NTIS No. PB-228-581 (September 1973).

70. Commonwealth of Massachusetts Metropolitan District Commission. "Cottage Farm Combined Sewer Detention and Chlorination Station, Cambridge, Massachusetts," U.S. EPA Report-EPA-600/2-77-046, NTIS No. B-263-292 (November 1976).

71. Glover, G.E. "High-Rate Disinfection of Combined Sewer Overflow, Combined Sewer Overflow Seminar Papers," U.S. EPA Report - 670/2-73-077, NTIS No. PB-231-836 (November 1973).

72. Tifft, E.C., P.E. Moffa and S.L. Richardson. "The enhancement of high-rate disinfection by the sequential addition of chlorine and chlorine dioxide," in *Proceedings of the Workshop on Microorganisms in Urban Stormwater,* U.S. EPA Report-EPA-600-2-76-244, NTIS No. PB-263-030 (November 1976), p. 96.

73. Drehwing, F.J., A.J. Oliver, D.A. MacArthur and P.E. Moffa. "Disinfection/Treatment of Combined Sewer Overflows," U.S. EPA Report-EPA-600/2-79-134, NTIS No. PB-80-113-459 (August 1979).

74. Moffa, P.E., E.C. Tifft, S.L. Richardson and J.E. Smith. "Bench-Scale High-Rate Disinfection of Combined Sewer Overflows with Chlorine and Chlorine Dioxide," U.S. EPA Report-EPA-670/2-75-021, NTIS No. PB-242-296 (April 1975).

75. Peicuch, P.J. "The chlorine controversy," *J. Water Pollut. Control Fed.* 46(12):2637 (1974).

76. Zillich, J.A. "Toxicity of combined chlorine residuals to freshwater fish," *J. Water Pollut. Control Fed.* 44(2):212 (1972).

77. Brungs, W.A. "Effects of residual chlorine on aquatic life," *J. Water Pollut. Control Fed.* 45(10):2180 (1973).

78. Ward, R.W., and G.M. DeGraeve. "Residual toxicity of several disinfectants in domestic wastewater," *J. Water Pollut. Control Fed.* 50:46 (1978).

79. Roeber, J.A., and F.M. Hoot. "Ultraviolet Disinfection of Activated Sludge Effluent Discharging to Shellfish Waters," U.S. EPA Report-EPA-600/2-75-060 (December 1975).

80. Oliver, B.G., and J.H. Carey. "Ultraviolet disinfection: an alternative to chlorination," *J. Water Pollut. Control Fed.* 48(11):2619 (1976).

81. Severin, B.F. "Disinfection of municipal wastewater effluents with ultraviolet light," *J. Water Pollut. Control Fed.* 52(7):2007 (1980).

82. Petrasek, A.C., H.W. Wolf, S.E. Esmond and D.C. Andrews. "Ultraviolet Disinfection of Municipal Wastewater Effluents," U.S. EPA Report-EPA-600/2-80-102 (August 1980).

83. Venosa, A.D. "Current state-of-the-art of wastewater disinfection," *J. Water Pollut. Control Fed.* 55(5):457 (1983).

84. Scheible, O.K., and C.D. Bassell. "Ultraviolet Disinfection of a Secondary Wastewater Treatment Plant Effluent," U.S. EPA Report-EPA-600/2-81-152, NTIS No. PB-81-242125 (August 1981).

85. Scheible, O.K., and A. Forndran. "UV Disinfection of Secondary Effluent and CSO Wastewaters," U.S. EPA Report-EPA-600/2-86-005, NTIS No. PB86-145182 (December 1985).

86. Scheible, O.K., R. Smith and W. Cairns. "The application of ultraviolet disinfection to combined sewer overflows," prepared for the New York City Department of Environmental Protection (June 1991).

87. Whitby, G.E., and F. Engler. "A preliminary study to determine the feasability of medium pressure lamps for disinfecting low quality wastewaters," Prepared for the Research Management Office, Ontario Ministry of the Environment, RAC Project No. 380C, Toronto, Ontario (September 1988).

88. Bank, H.L., J. John, M.K. Schmehl and R.J. Dratch. "Bactericidal effectiveness of modulated UV light," *Appl. Environ. Microbiol.* 56(12):3888 (1990).

STORMWATER RUNOFF AND FLOOD CONTROL

15 TELEMETRIC MONITORING OF WATER LEVELS FOR A DRAINAGE SYSTEM IN SINGAPORE

15.1 INTRODUCTION

In 1990, the Drainage Department of the Ministry of the Environment installed a computerized telemetric system, costing $230,000, for monitoring one of the country's most important drainage facilities, the Bukit Timah-Kallang River System. The system provides information on the water levels and rainfall conditions at the monitored areas. In addition, it has the capability for generating automated flood-warning messages.

The physical layout of the system is depicted in Figure 1. It consists of a central monitoring station linked to seven water-level monitoring stations and two rainfall monitoring stations within the Bukit Timah-Kallang catchments, and telex terminals (at up to four selected destinations) for receiving flood-warning messages.

15.2 SYSTEM DESCRIPTION

Figure 2a shows the network configuration of the telemetric monitoring system. The system components are listed in Table 1.

Linkage between the central monitoring station and the remote stations are through dedicated telecommunication lines of the public telephone system.

The salient details and functions of the principal system components are described as follows.

0-87371-805-4/93/$0.00+$.50
© 1993 by Lewis Publishers, Inc.

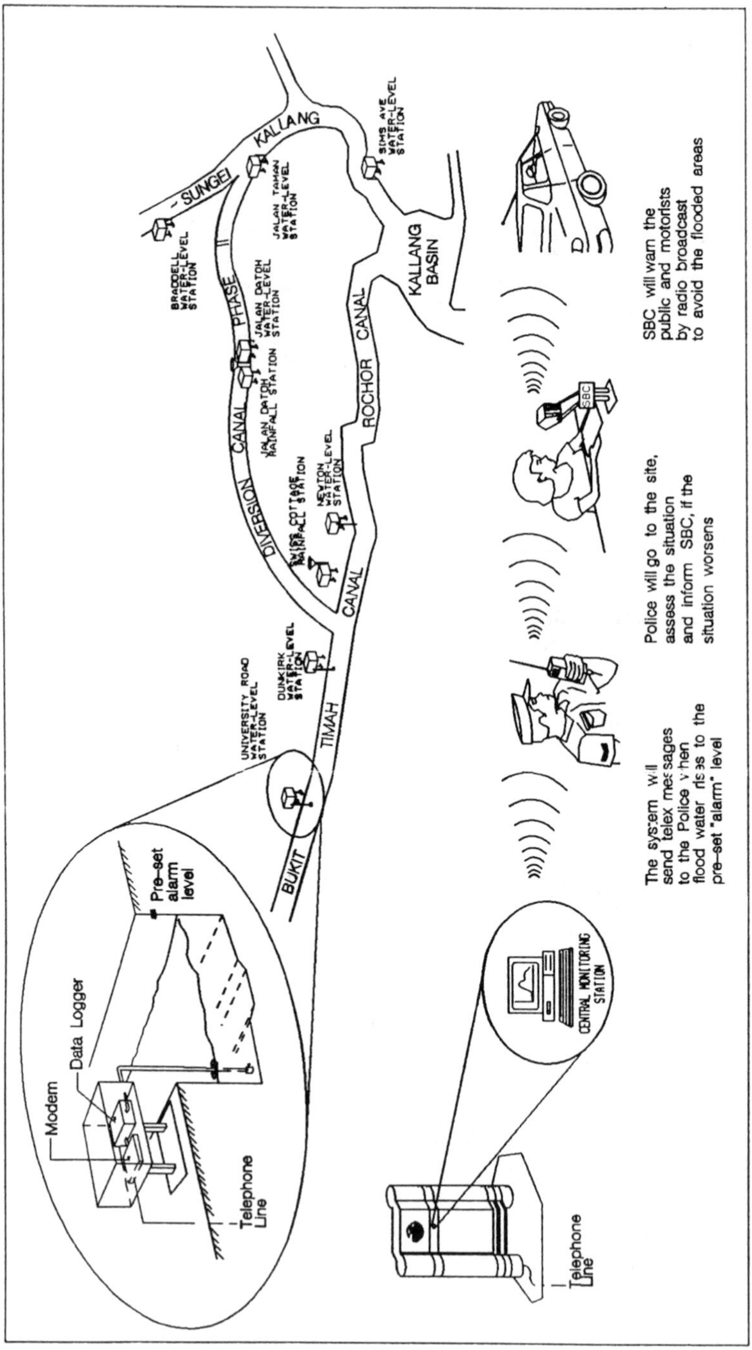

Figure 1. Telemetric monitoring system for the Bukit Timah-Kallang River System.

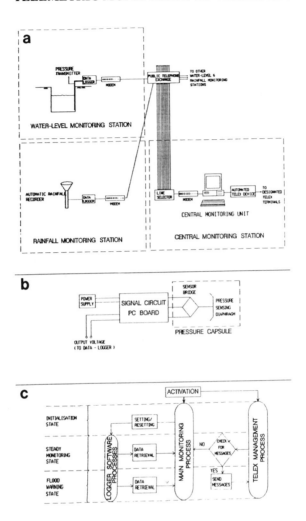

Figure 2.   **(a) Network configuration. (b) Schematic functional diagram of pressure transmitter. (c) System processes.**

15.2.1 Central Monitoring Unit (CMU)

The CMU is the system's host computer located at the central monitoring station. It is an 80286-based microcomputer driven by the DOS operating system. The computer is equipped with a VGA monitor (as the system's display unit), a 30-megabyte hard disk and other auxillary peripherals. Residing in the CMU are two main software systems: the main monitoring system which retrieves real-time information from the remote monitoring stations and generates flood-warning messages under certain predefined conditions; and the

TABLE 1. System Components

Station(s)	Components
1 At the central monitoring station	(a) The central monitoring unit (CMU)
	(b) The line selector
	(c) The automated telex device
	(d) A telecommunication modem
2 At each of the seven water-level monitoring stations	(a) A data logger
	(b) A pressure transmitter
	(c) A telecommunication modem
3 At each of the two rainfall monitoring stations	(a) A data logger
	(b) An automatic rainfall recorder
	(c) A telecommunication modem
4 Others	(a) Up to four telex terminals at the designated agencies (including the Police Department) to receive flood-warning messages
	(b) Portable lap-top interrogater

automated telex subsystem which manages and transmits the generated flood-warning messages to predefined telex terminals.

15.2.2 Line Selector

This is a 16-way relay output module which is connected to the CMU through a modem. The dedicated telecommunication lines from the remote monitoring stations are fanned into the line selector. By switching on the contact with the appropriate telecommunication line, the line selector links the CMU to a selected monitoring station, thereby allowing the CMU to communicate with only one remote station at a time without interference.

15.2.3 Data Loggers

The data logger is a microprocessor-based data acquisition unit at each remote monitoring station. The data logger used in this system has 32 kb of ROM (read-only memory) and 24 kb of RAM (read-and-write memory) with a data storage capacity of 10,800 points. All operating software is built into the ROM. Supervised from the CMU, the data logger converts the analog signals (in the form of measured voltages) from both the pressure transmitters of the water-level monitoring stations and the rainfall recorders of the rainfall monitoring stations, into standard ASCII formats suitable for input to the CMU.

Operation of the data logger and retrieval are normally controlled through commands from the CMU. In the event of telecommunication failure, its control and data retrieval may be carried out on site using a portable interrogator. The data logger is provided with a RS232 serial interface for remote communication at 1200 bauds through the public telephone network and for on-site communication at up to 4800 bauds.

The data is stored serially in the logger as a FIFO (first-in-first-out) stack. The CMU scans and retrieves the stored data at preset intervals or as commanded by the system operator, and erases the data stack after successful retrieval.

15.2.4 Pressure Transmitters

The pressure transmitter is the electronic water-level sensor at each of the water-level monitoring stations. It is essentially a variable voltage output device whose output is proportional to the pressure of water in the drainage channel. The pressure transmitter, shown schematically in Figure 2b, comprises two main elements: the pressure capsule, which consists mainly of a sensor diaphragm and a sensor bridge generating differential electrical voltages in response to applied water pressures; and the signal conditioning PC (printed circuit) board, which consists of the circuitry for regulating and amplifying the differential voltages from the pressure capsule.

The pressure capsule is the "sensing probe" of the transmitter in that it performs the function of tranducing the detected changes in water pressure into output voltages. At zero water pressure, the output voltage is adjusted to zero with the aid of a "zero adjustment potentiometer" in the signal conditioning PC board. As water pressure is applied, the sensor diaphragm receives and transmits the water pressure to a strain-sensitive element connected to the sensor bridge (a Wheatstone Bridge circuit). The strain generated in the element causes an electrical imbalance to the sensor bridge, thereby generating a small differential voltage output (in millivolts). The signal conditioning PC board raises and adjusts the output voltage to the calibrated level. A span potentiometer in this PC board performs the calibration by adjusting the final output voltage to 5 volts DC at a full-scale pressure range of 12 m of water. This output voltage is recorded in the data logger.

15.2.5 Automated Rainfall Recorders

The rainfall recorder is a tipping-bucket rain gauge connected to the data logger. One filling of the tipping bucket of this rain gauge corresponds to 0.1 mm of rainfall. The bucket tips and empties its contents automatically after it is filled, closing a reed contact to produce an electrical impulse recorded in the data logger.

15.2.6 Automated Telex Device

This is a dedicated microprocessor which enables the CMU to function as an automated telex-transmitting terminal. The device has 32 kb of RAM acting as a buffer to store flood-warning messages generated by the CMU. It is driven by the automated telex subsystem software (known as "telexware") that resides in the CMU. Under certain conditions predefined by the system operator, the telexware is activated to send flood-warning messages to designated telex terminals.

15.3 SYSTEM OPERATION

The operation of the telemetric monitoring system involves distributed, parallel processing under the supervision of the CMU, and asychronous wide-area communication between the parallel software processes. Three main software processes are in execution during the system's operation:

1. The main monitoring process residing at the CMU — this process runs in the interactive mode and controls the operation of the entire system.
2. The telex management process, also residing at the CMU — this process runs in the background mode and may be activated by the main monitoring process to send flood-warning messages.
3. The logger software processes residing at the site loggers — these are distributed parallel processes that run under the supervision of, and communicate asynchronously with, the main monitoring process of the CMU.

Figure 2c outlines the salient process procedures and the interractions among these processes. After the main monitoring and telex management processes have been activated in the CMU, the system may be in one of the three operational states that follow.

15.2.1 Initialization State

Under operator's instructions at the CMU, the main monitoring process sends remote commands to clear the RAM storage of the field loggers and sets up (or resets) the logger's mode of data scanning and recording. One important purpose of this initialization procedure is to synchronize the loggers' clocks with the CMU's clock. This is a subtle but important aspect of the system operation which will be discussed in a later section.

15.3.2 Steady Monitoring State

After the initialization state, the system enters the steady monitoring state during which the field loggers continously scan and log the water level and rainfall data, while the main monitoring process periodically retrieves the logged data from the remote monitoring stations, either at pre-defined intervals or under operator's instructions at the CMU. The programmed retrieval procedure (referred to as "polling") is performed by the following steps:

1. The main monitoring process initiates and establishes a software "handshake" (interactive device control) with the logger software process of the first remote monitoring station, and thereafter commands the remote data logger to transmit data.
2. The remote data logger transmits through its modem all freshly logged data (data logged since the previous retrieval) in bit-serial sequence to the CMU.
3. The CMU stores the retrieved data into labeled files under DDMMYY.STN (day, month, year and station number) nomenclatures.
4, The CMU sends an end-of-data code ("CLAST") to the remote logger.
5. The remote logger clears its stored data after receiving the "CLAST" code, and "echos" the code back to the CMU.
6. The CMU closes communication with the first station, and establishes a "handshake" with the second station.
7. Data retrieval continues until all activated remote stations have been polled (stations may be deactivated by the operators at the CMU before polling begins, in which case such stations would continue their data logging processes independent of CMU's interference).

15.3.3 Flood-Warning State

The main monitoring process generates flood-warning messages when the water level at any monitored site reaches a predefined "alarm level". Such warning messages will continue to be generated until the water level recedes below a predefined "clear level". The background telex management process "picks up" and transfers the messages into the automated telex device for transmission to the designated remote telex terminals.

15.4 SYSTEM TUNING

From the point of completing the initial set-up of the system, it took about three months of trial runs and customization (referred to here as "system tuning") before the system could be successfully put into operation. The system-tuning issues encountered may be broadly classified into the two categories that follow.

15.4.1 Design Tuning

The design tuning issues were mainly associated with the system's need for asynchronous wide-area communication among devices with different clock rates. Two critical data-corruption problems were detected and resolved, as follows:

Station Cross-Over Problem

Situation: The data retrieved from the previously polled station were written into the record file of the subsequently polled station.

Analysis: A data signal traveling at 300,000 km/sec (assumed to be at the speed of light) would take 10,000 ns (nanoseconds) to complete a telecommunication path of 3 km from the remote monitoring station to the CMU. This is about 100 times the CMU's clock periods (with a clock rate of 10 MHz) and, significantly, an order of 20 times the logger's clock rate of about 500 nsec. Depending on the telecommunication path taken (as may be affected by telecommunication traffic and interchange control), it is possible for the first signal to be delayed (e.g., by 10 to 20%) thereby reaching the CMU many clock periods later than some subsequently sent signals. Whenever the earlier signals from a remote station arrived at the CMU after the end-of-data code (a later signal) is received, the above-mentioned "station cross-over" problem would ensue.

Solution: This was solved by introducing a sufficiently "idle-wait" time between the receipt of the last data signal and the final transmission of the end-of-data code.

Day Cross-Over Problem

Situation: Day-one data are written into the Day-two's record file for the same station, and other similar problems.

Analysis: After initialization of the loggers and synchonization of the loggers' clocks with the CMU's clock, the loggers' clocks gradually "drift" out of synchronization. The CMU necessarily schedules data retrieval based on its own clock, whereas the loggers have to store and "time-label" the scanned data based on their respective clocks. This can result in a large discrepancy between the data's time-labels and the CMU's time at the crossing of day (e.g., between 2359 hours at a logger and .0001 hours at the CMU), giving rise to the abovementioned "day cross-over" problem.

Solution: The monitoring software was improved to activate two consecutive pollings, one each immediately before and after midnight, thereby minimizing the band of data affected by the "time-drift". The "drifts" are also minimized operationally by initializing the loggers fortnightly.

15.4.2 Operational Turning

The operational tuning issues were, on the other hand, mainly related to the following operational parameters:

- polling intervals — time periods between consecutive retrievals of loggers' data to the CMU
- data density — the number of recorded data points per hour

TABLE 2. Effect of Paramter Variation on System Function

System's Operational Characteristics	Definitions	Effects of Variations of the Following Factors:	
		Polling Interval	Data Density
Quality of records	Usefulness of data for monitoring & analysis	Not affected (within normal operational range)	Quality improves with data density
Polling time	Time taken for data retrieval during each poll	Increases with lengthened polling interval	Significantly lengthened by increase in data density
Alertness of system	Timeliness of flood warning & information	Shortening polling interval heightens alertness	Slightly reduced by increased data density (within normal operational range)

These parameters affect the system's functions as summarized in Table 2.

Figure 3 shows the variations of polling time with variations to number of stations and the polling intervals (Figure 3a), and the monitored results of the drainage system's reaction to typical rainstorms (Figure 3b). From the graphs and the above table, the following deductions have been made:

- Relatively high-quality records (dense data points) are needed to capture the sharp variations of water levels.
- Short polling intervals are needed to achieve a satisfactory level of system alertness, owing to the relatively short time available for flood warning.
- There exists the possibility of shortening the polling time by reducing the polling intervals, but within certain optimum limits (as some overheads are required for establishing communication handshakes even if no data are available for transfer).

An important operational point to note is that once the polling procedure is activated, it captivates the CMU and suspends all other CMU's processes. Shortening the polling interval to below the threshold polling time will, for instance, freeze the CMU into perpetual polling of data and completely disable (rather than enhance) the system alertness.

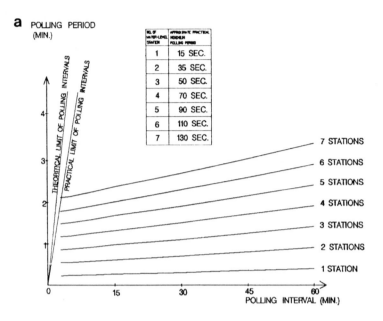

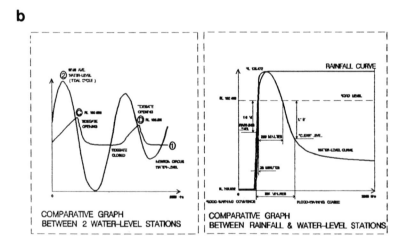

Figure 3. **(a) Limitation of polling intervals. (b) Monitored results of drainage system's reaction to a rainstorm.**

Decreasing the data density (within limits) will also improve the system's alertness owing to the corresponding reduction in data loads, but this will degrade the data quality as discussed above. Furthermore, extreme reduction of the data density may cause the CMU to be "busy polling", without obtaining fresh data, at times. Owing to variations of time required for establishing telecommunication handshakes and data transmission, and the occasional occurrences of corrupted data points, the author's recommendation is that there should be at least four data points within each polling interval.

The system has now been set (under normal conditions) to perform half-hourly polls, with about 15 data points per poll. Once a significant rise of water level (currently set at 1 m/h) is detected, it will automatically decrease the polling interval to 15 min with 7 to 8 data points per poll, thereby stepping up the system's alertness for flood warning.

15.5 SYSTEM'S BENEFITS FOR STORMWATER MANAGEMENT

This telemetric monitoring system is a useful stormwater management tool for the Bukit Timah-Kallang River catchment, with the following main applications.

15.5.1 Real-Time Monitoring of Drainage System

The system enables remote, real-time monitoring of the main Bukit Timah-Kallang River System. Engineers of the Drainage Department can assess, quickly and accurately, the overall conditions of the drainage system. As an example, the results of monitoring shown in Figure 3b provide a ready assessment of the operation of the tidegate at Rochor Canal (the downstream reach of Bukit Timah Canal). It may be deduced from a comparison between Curve 1 (water level variation at Newton Circus upstream of the tidegate) and Curve 2 (tidal fluctuation near Kallang Basin downstream of the tidegate) that the tidegate had been correctly operated such that it was closed at times of high tides and opened timely (at points "C" and "D") to release the storm flows.

15.5.2 Advanced Flood-Warning

The capability of the system to generate automated, on-line flood-warning messages has been discussed at length. Through successful implementation of an operational plan set up by Drainage Department with the agencies involved (pictorially presented on the lower part of Figure 1, the system provides timely flood-warning information to be broadcasted to motorists and for the police to initiate early actions to divert traffic from the flooded roads in the monitored areas.

15.5.3 Hydraulic Modeling Studies for the Drainage System

The system provides the essential data for drainage master-plan review to cater for long-term flood protection of the Bukit Timah-Kallang catchments, as this central core of Singapore will become more intensely developed into the 21st century. The collated data, in standard ASCII digital format, can be electronically extracted for computer modeling studies. An applied research project has already been initiated by the Drainage Department in this direction.

15.6 CONCLUSIONS

The installation of the telemetric monitoring system for the Bukit Timah-Kallang River System has involved the selection and appropriate customization of available technology, to cater for the monitoring system's intended hydrologic and operational functions, and to suit the specific characteristics of the drainage system. Planning and implementation of this system called for the application of computer technology with stormwater management know-how.

The drainage Department's experience has found the telemetric monitoring system to be a valuable stormwater manangement tool.

REFERENCES

1. Hayes J.P. *Computer Architecture and Organization* (New York: McGraw-Hill International Editions, 1988).
2. Deitel, H.M. *Operating Systems* (Reading, MA: Addison-Wesley Publishing Co., 1990).
3. Campbell, J. "The RS-232 Solution," Sybex/Tech Asian Editions (1988).

16 CONCEPTS FOR FLOOD CONTROL IN HIGHLY URBANIZED AREAS

16.1 INTRODUCTION

Water is the basic requirement for living as well as it may be a threat to lives. This natural change in condition has been aggravated by urbanization. Although the disadvantages of urbanization for the ecosystem and human well-being are recognized, people continue to migrate from rural to urban areas (Figure 1). More than 50% of the world's population, which will increase from 5.4 billion people today to approximately 8 billion people in the year 2000, will concentrate in urban areas. Also, in industrialized countries, such as Germany, urbanization still increases. In Germany today 12% of the total area is urbanized. Urban areas continue to increase by 400 km^2 each year; and approximately $^1/_3$ of urban surfaces are impervious. To understand and control the interaction of urbanization and stormwater runoff a comprehensive view of the urban water cycle and the integrated water management of urban and surrounding areas is necessary.

16.2 EFFECTS OF URBANIZATION ON THE HYDROLOGICAL REGIME AND ON ECOLOGY

Urban land uses have a great impact on water quantity and quality aspects of the hydrological regime.

16.2.1 Increase in the Incidents of Floods

Fast concentration of runoff on impervious surfaces and hydraulic improvements in the form of gutters, storm sewers, and drains result in quickly-peaking high runoff rates. This tendency is supported by the straightening, deepening,

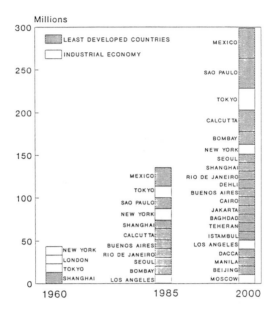

Figure 1.   Increase of urban areas with more than 10 million people.

and lining of natural river beds and channels within urban areas as well as in the upstream river basin.

Rao et al.[1] found that an increase in the areal imperviousness from zero to 40% would approximately halve the time to peak discharge and increase its magnitude by 90%. The increase in surface runoff from urban areas may cause local flooding and flooding in downstream areas, thus endangering human lives and wildlife, disrupting ecosystems and human activities, and damaging houses and other properties. This situation is substantially worsened by deforestation in upstream catchments.

16.2.2 Reduction of Base Flows and Groundwater Recharge

Impregnating surfaces by impervious elements such as rooftops, streets, sidewalks, and parking lots, and consolidating topsoils by land use greatly reduce rainfall abstractions, which in nature are achieved by vegetation, depressions, and infiltration into the ground. Impervious surfaces result in immediate runoff and reduce groundwater recharge, which again lowers groundwater tables. At the same time, stream base flows are reduced.

Calculations for a catchment in California indicated that complete urbanization increased runoff volumes 2.3 times compared to the predevelopment state, while the stream base flow fell to 0.7 of its natural magnitude.[2] This problem is aggrevated by withdrawing groundwater for urban uses, leading to land subsidence. Consequently, urban drainage in cities like Bangkok or Houston becomes most difficult. Furthermore, runoff from some urban areas may be contaminated and pollutes groundwater, if infiltrated.

16.2.3 Increase in Pollution

The more that energy and resources are consumed in urban areas, the more wastes are produced and discharged. Construction activities increase soil erosion and the discharge of suspended solids into receiving waters. Dallaire[3] estimated erosion rates for disturbed urban areas as 28 ton/ha/year and 0.2 ton/ha/year for well-established urban areas. Concentrated human activities lead to the deposition of dust, dirt, and various pollutants on the catchment surfaces. Organic material, salt, and toxic heavy metals are eventually washed off by storm runoff and contribute to the pollution of receiving waters. If stormwater is infiltrated, groundwater may be polluted.

16.2.4 Impact on Receiving Water Ecology

Flood flows in natural catchments increase over a certain time span, allowing aquatic organisms to refuge into their protection spaces. Stormwater discharges to urban waters have an hydraulic and pollutional impact. The hydraulic shock loads occur suddenly and frequently. The bottom shear stress in rivers is immediately increased. Organisms drift downstream. Bottom material is eroded so that organisms hiding in interstitial spaces are flushed out. Sometimes the complete biocynosis is eliminated. In addition, moving stones and particles destroy organisms. Small urban waters especially suffer from the quickly peaking storm discharges. Stormwater discharged to receiving waters may reduce water quality even more than the discharges from sewage treatment facilities. Although many of the transported materials are of natural origin and relatively harmless when deposited on land, they become pollutants in receiving waters. Organic matters deplete oxygen. Oils, salt, and particles from car tires are especially toxic to aquatic life, either directly or by a cumulative effect. Solids and suspended matters clog interstitial spaces at the river bottom. Nutrients ultimately cause algae growth. Anaerobic decomposition of organic matters may lead to odor nuisance.

16.2.5 Climatological Changes

Large urban areas affect the local climate by increased air temperatures, reduced humidity, higher incidence of fog, and increased precipitation. Air temperatures in urban areas are commonly higher than in their environs. The wind velocity in urban areas is lower than in rural areas because of the obstacles caused by buildings which change the natural flow and turbulence of the air. The humidity of the air is also lower in urban areas because the rainwater is quickly removed from impervious surfaces.[4]

16.3 TECHNICAL CONCEPTS FOR FLOOD CONTROL IN URBAN AREAS

Concepts for flood control in urbanized areas should counterweight the effects of urbanization. Vegetation and soil cover contribute to the hydrologic cycle in the form of evapotranspiration and storage. These are lost through urbanization and must be replaced by technical measures. Therefore, the principles for urban drainage should be to:

1. limit impervious surfaces to a minimum
2. store and utilize stormwater to a maximum.
3. detain and treat storm runoff as much as possible
4. discharge only the unavoidable runoff into receiving waters
5. line receiving waters in a natural way

The technical means may be subdivided into structures for infiltration of stormwater into the ground and for stormwater storage and detention.

16.3.1 Infiltration of Stormwater

Investigations have shown that (for middle-European conditions) stormwater runoff from 20 to 30% of the usually impervious urban areas may be infiltrated into the ground. However, only about 5 to 10% of paved surfaces may be directly replaced by pervious covers. In other words, the infiltration potential is much higher than the potential for the removal of impervious areas.

Direct infiltration of runoff from impervious areas such as roofs, paved streets, or sidewalks may be enhanced, especially in residential areas. The use of porous road and parking lot covers has also been considered. On the other hand, if the runoff from streets and parking lots is expected to be heavily polluted, its infiltration into the ground should be avoided. Once a groundwater aquifer has been contaminated, the removal of pollutants is physically and economically unfeasable.

16.3.2 Surface-Ponding and Reuse of Stormwater

Stormwater runoff may be temporarily detained by storage on catchment surfaces. The areas used for this purpose may include parking lots, flat rooftops, playgrounds, public gardens, parks, and special flood areas. Open storage facilities in urban areas should be designed so that they flood only under severe storm conditions. The retained water is released slowly after the storm or infiltrated into the ground. An example for this is the dual-drainage concept of several municipalities in Southern Ontario, Canada. At rare rain events, runoff exceeding the capacity of street inlets is conveyed by streets to detention areas in parks.

Under appropriate climatological conditions, storage and reuse of stormwater for irrigation or other subpotable water supplies should be considered. Large stormwater tanks up to 3000 m^3 in volume have been installed for water for fire extinction. In Singapore, storm runoff from residential areas is collected, stored, and even treated for potable use.

16.3.3 Major and Minor Drainage Systems

The minor drainage system comprises swales, street gutters, catch basins, storm sewers, and surface and subsurface detention facilities. It conveys runoff from frequent storms with return periods up to the design period, namely from 1 to 10 years. The minor systems primarily reduce the frequency of inconvenience caused by stormwater ponding to both pedestrians and motorists. The consequences of the failure of the minor drainage systems are often insignificant, provided that there is a properly functioning major drainage system. The major drainage system comprises natural streams and valleys as well as manmade drainage elements, such as channel and flood detention facilities. This system should accomodate runoff from infrequent storms with long return periods up to 100 years or more. Appropriate design, construction, and maintenance of major systems greatly reduce the risk of loss of life and the probability of damage caused by flooding. It should be emphasized that the major drainage system exists in nature regardless of whether or not it is identified and preserved during the urban development.

The linkage of the design of the minor and major drainage systems has been a major problem in the past. Due to the lack of appropriate design methods in the past both systems have been designed separately. This frequently leads to conditions where the major system is overloaded while the minor system still shows empty storage capacities and vice versa.

16.3.4 Flow Diversion and Flood Detention

Obviously floods with a recurrence of, for instance 100 years or more, cannot be defeated without additional measures. It is still necessary to provide large detention basins and/or diversions designed for the desired flood protection. To this extent urban drainage planning obviously is linked to the capacity of the receiving-water system. Therefore urban drainage plans should be part of a watershed plan and preferably be prepared at an early stage of the watershed development.

16.4 MODEL CONCEPT FOR INTERACTIVE DESIGN OF MAJOR AND MINOR DRAINAGE SYSTEMS

For planning and design of major and minor drainage systems it is impossible to derive the required design data directly from measurements of the past because the surface structures, due to urban development, continuously change. To calculate the interaction of natural catchment and urban runoff a deterministic modeling approach seems to be most appropriate. If all components of the drainage system that determine translation and retention of runoff are considered separately in the model, any existing and anticipated drainage condition may be represented. It is important, however, that for known conditions, the model is verified.

16.4.1 Model Structure

To secure a flexible model concept the model consists of the following components:

1. rainfall selection and loading characteristics
2. rainfall abstractions
3. runoff concentration
4. flow detention and control
5. flood routing

16.4.2 Rainfall Selection and Loading Characteristics

The model concept allows for input of natural and synthetic rainfall events. The maximum rainfall duration may be 72 h. For synthetic storms, three options frequently used in Germany, are available. However, any model storm concept may be used as input.

16.4.3 Rainfall Abstractions

For the impervious part of urbanized catchments effective rainfall is calculated by subtracting an initial loss from the original rainfall. In reality runoff starts before all depressions are filled. Therefore the impervious area is divided into three equal portions. In one portion, only $1/3$ of the initial loss, in the second portion, the full initial loss, and in the third portion, $5/3$ of the initial loss is applied. For the pervious part of urbanized catchments, the SCS concept is adapted. This concept is also used for natural catchments. However, for middle-European conditions, it was found that instead of an initial loss of 20%, a value of 5% of the potential infiltration capacity of the soil fits measurements quite well.

16.4.4 Runoff Concentration

For urbanized catchments, runoff concentration is calculated using a parallel series of linear storage reservoirs (Figure 2). The parameters n_1 and n_2 express the number of storages in each series. Comparison with measurements resulted in a number of three storages each. The storage coefficients k_1 and k_2 are dependent on the surface characteristics.

For runoff concentrations on natural catchments, three parallel series of linear storage reservoirs were chosen. The first series represents the runoff from rather impervious portions or from areas close to receiving waters, which result in

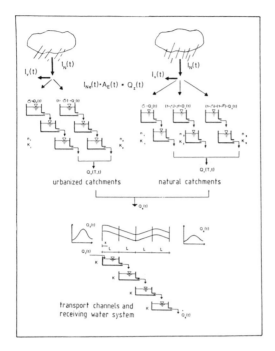

Figure 2.   Integrated rainfall runoff model for major and minor drainage systems.

direct discharge. The second cascade represents the slower and somewhat detended runoff from the catchment. The third series of linear reservoirs represents the interflow in subsurface soil layers.

The parameters n_1, n_2 and n_3 were chosen (on the basis of comparison with measurements) to be two. The storage coefficients k_1, k_2 and k_3 again depend on the catchment characteristics and must be calibrated.

16.4.5 Flow Detention and Control

Within the minor drainage system the structures summarized in Figure 3a are considered. To represent surcharge conditions in the local drainage networks a hypothetical detention facility may be added reflecting the anticipated surcharge volume. Within the major drainage system, which usually corresponds to receiving waters, the flood control structures of Figure 3b usually occur. The inflow to the structures is stored and diverted according to the values provided for storage volume and throttle flow.

16.4.6 Flood Routing

Flood routing in the major and minor drainage system is calculated according to the Kalinin-Miljukov method which is based on a series of linear storage

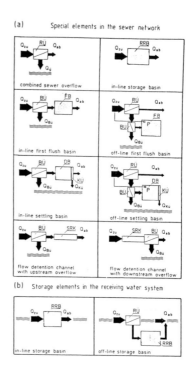

Figure 3. Overflow and storage facilities considered within (a) minor and (b) major systems.

reservoirs using the same storage constants within a certain flow segment (Figure 2). Detention and routing segments may be composed similar to reality, i.e., detention facilities may be located at any point in the system.

16.5 EFFECTS OF DIFFERENT STRATEGIES ON FLOOD PEAKS

The potential of structural and nonstructural measures taken within an urbanized catchment were investigated for their effect on flood peaks. The following discussion contains some results from studies done for the Emschergenossenschaft in Essen. This authority is responsible for the water management in the highly urbanized and industrialized basin of the Emscher River. Two and one half million people live in the river basin, which is 860 km^2 in size. The population equivalent for sewage treatment amounts to five million. At present an open system is being operated that has solved the difficult problem of waste water collection and storm drainage under ongoing land subsidence due to past coal mining. A major objective for a new stormwater drainage concept is

**Table 1. Individual Abatement Measures Investigated
for Their Effect on Flood Reduction[5]**

Description of Measure	Abbreviation
Decentral storage on surfaces for stormwater reuse (5 mm/m^2 = 50 m^3/ha$_{red}$)	Renat.
Reduction of impervious areas and stormwater infiltration (5 %)	Red. imp.
Retention of combined flows (30.5 m^3/ha$_{red}$)	Ret. 30.5
Retention and additional detention of combined flows (60 m^3/ha$_{red}$ and 120 m^3/ha$_{red}$ resp.)	Det. 60 Det. 120
Renaturation of river profiles	Renat.
Flood flow detention (400 m^3/ha$_{red}$)	Flood det.

to halve flood peaks at the river mouth. This objective should facilitate integration of the diked Emscher River into the urban environment.

To show the effect of individual abatement measures on peak runoff the following were investigated: (1) decentral stormwater infiltration equivalent to a reduction of imperviousness by 5%; (2) decentral storage of storm runoff on surfaces for stormwater reuse equivalent to 5 mm/m^2; (3) semicentral detention of storm and combined flows at a storage rate of 30.5 m^3/ha$_{red}$ and at extended rates of 60 m^3/ha$_{red}$ and 120 m^3/ha$_{red}$, respectively; and (4) landscaping and widening of river profiles as well as flood flow detention at rates of 400 m^3/ha$_{red}$. Table 1 summarizes the individual abatement measures investigated.

With the runoff model described above, the effects of the individual measures were calculated for model storms of return intervals of 2, 5, 10, 20, 50, 100, and 200 years. It was found that at the mouth of the Emscher River the critical flood occurred with storms of a duration of 12 h. For runoff calculations, the Emscher system was divided into 246 urbanized subcatchments, 7 natural subcatchments, 415 open transport elements, 76 underground transport elements, 246 detention basins for combined flows, 246 detention basins at combined overflows, 86 pumping stations, and 51 flood detention basins in the receiving-water system, whereof 8 are existing. The high discretization of the Emscher area for calculation purposes was necessary to show the effect of the individual drainage elements on flood reduction in the tributaries.

Figure 4 shows that for high return intervals all decentral measures show little effect on flood reduction. This is especially true for infiltration and decentral storage at the considered rates. Up to return intervals of 5 to 10 years, stormwater detention within the minor drainage system is quite effective. However, for high return intervals of storm loadings, the effect diminishes to almost zero. On the other hand, large detention facilities within the major drainage systems reduce flow peaks significantly.

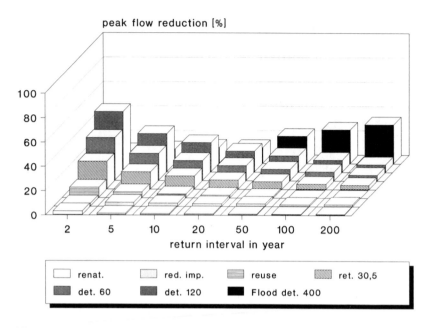

peak flow reduction [%]

Figure 4.   Reduction of flood peaks by different nonstructural and structural measures.[5]

Further, the combined effect of different measures (Table 2) was studied. All combinations not containing flood detention facilities in the major system have a limited effect at all events with high return intervals. Combinations with detention in the major system and with operational control of the available storages would provide a reduction of flood flows in the magnitude of 60% for all events up to return intervals of 200 years. Figure 5 compares the effectiveness of different combinations. The numbers used in Figure 5 correspond to Table 2.

16.6 CONCLUSIONS AND RECOMMENDATIONS

Urban areas will grow at an extent never experienced before. Appropriate sanitation and flood control for these areas can only be achieved if the rainfall runoff process already is considered in the planning phase as one unique system. A planning procedure is suggested that comprises hydrologic calculations for minor and major system components. With this model concept, a number of nonstructural and structural flood abatement measures were investigated. It was found that stormwater reuse and stormwater infiltration, if applied within urban areas, only had a marginal effect on the reduction of flood peaks. Some effect was achieved with semicentral detention corresponding to a storage volume of 30 m³/ha impervious area. Flood detention of a magnitude of 400 m³/ha

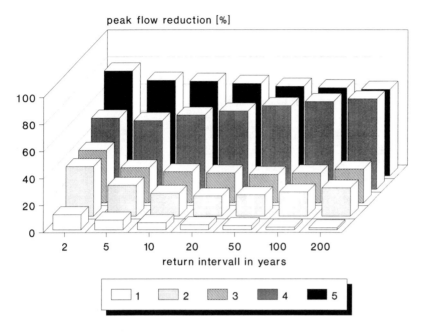

Figure 5.   Combination of drainage measures and their effect on flood peak reduction.[5]

Table 2. Combined Measures Investigated for Their Effect on Flood Reduction[5]

Description of measures	Abbreviation
Combined measures (no. 1) (reuse 50 m³/ha + red. imp. 5%)	R + I
Combined measures (no. 2) reuse + red. imp. + ret. 30.5 m³/ha + renat.	R + I + R 30.5 + N
Combined measures (no. 3) reuse + red. imp. + det. 120 m³/ha + renat.	R + I + D 120 + N
Combined measures (no. 4) reuse + red. imp. + ret. 30.5 m³/ha + ren. + fl. det.	R + I + R 30.5 +N + f D
Combined measures (no. 5) as before and additional operation control	Add. oper. ctl.

impervious area finally reduced flood peaks significantly. It is suggested that for urban catchments the interrelated effects of all possible measures be studied and the results incorporated into developmental policies.

ACKNOWLEDGMENTS

The author very much acknowledges the permission of the Emscher-genossenschaft to discuss some findings of studies towards a new drainage concept for the Emscher River basin. In specific, the financial support of the Emschergenossenschaft to the University of Essen for a number of investigations is highly appreciated.

REFERENCES

1. Rao, R.A. et al. "Conceptual hydrologic models for urbanizing basins," *J. Hyd. Div. ASCE,* 98:HY7 (1972).
2. James, L.D. "Using a computer to estimate effects of urban development on flood peaks," *Water Resour. Res.* 1(2): 223 (1965).
3. Dallaire, G. "Controlling erosion and sedimentation at construction sites," *Civil Eng.* 47 (10):73 (1976).
4. Geiger, W. F., J. Marsalek, W. J. Rawls and F. C. Zuidema, Eds. "Manual on drainage in urbanized areas. I. Planning and design of drainage systems," A contribution to the International Hydrological Programme, United Nations Educational, Scientific and Cultural Organization, Paris (1987).
5. Emschergenossenschaft. "Reduzierung von Hochwasserabflüssen in den Wasserläufen des Emscher-Gebietes," Internal report, Essen (1991).

17 USE OF TAPERED INLET TO INCREASE CULVERT CAPACITY

17.1 INTRODUCTION

Many existing stormwater drainage systems in Sydney have been in use for decades and some of them are now operating at their full capacities. Flooding occurs frequently, causing severe damage to surrounding properties. Local government councils have to allocate a substantial amount of their budgets to drainage improvement works. In the following sections of this chapter several important factors which contribute to poor performances of old drainage systems are listed.

17.1.1 Aging

After years of service, pipe settlement, leaky joints, erosion of pipe wall, blockages of inlets or outlets, accumulation of debris and penetration of tree roots restrict the drainage capacity. These problems have to be checked and rectified frequently.

17.1.2 Inaccurate Hydrological Data

Older systems were designed according to British design procedures using hydrological data which might not be applicable to Australian conditions. The rainfall intensities used earlier in some areas were found to be totally inadequate and resulted in failures of the drainage systems. More reliable hydrological data have now been collected and design information has been regularly updated in *Australian Rainfall and Runoff.*[1]

0-87371-805-4/93/$0.00+$.50
© 1993 by Lewis Publishers, Inc.

271

17.1.3 Poor Design Philosophy

The use of over-simplified design procedure such as the Rational Method does not give an accurate estimate of runoff because the method does not properly model the rainfall-runoff process. The effects of catchment characteristics on the response of the catchment to the rainfall are not really simulated. More advanced computer methods such as SWMM, ILSAX, RORB, RAFTS, MOUSE, etc., are now widely used.

17.1.4 Urbanization

Residential development in Australia has been mostly low-density housing, typically with a single story house on a block of land of around 1000 m^2. Rapid population growth results in subdivision of the blocks and redevelopment into medium-density housing (duplexes or townhouses) and high-density housing (flats or home units in multi-story buildings), particularly in the inner city suburbs. The associated increase in impervious areas, and hence surface runoff, often overloads the existing systems. Many councils now demand that property developers must provide on-site detention storages and control site discharge into the public drainage systems.

17.1.5 Poor Hydraulic Performance

Poorly designed inlets, junction pits, sudden change of flow direction, flow separation around sharp corners, and constriction of flow areas significantly increase energy losses and hence decrease drainage capacity. These problems can be improved by streamlining the flow. Hydraulic modeling is often used to evaluate the effectiveness of various improvement options.

Old drainage systems are usually upgraded by enlargement of the culverts or addition of parallel lines. But these options may not always be feasible because of cost or site constraint. In many cases, the capacity of a drainage system can be increased by simply adding a tapered inlet to streamline the flow. Keeping the barrel to a steep slope ensures that the culvert will operate under inlet control with critical depth and minimum specific energy occurring at the entrance and maximum flow rate is obtained under this condition. Rossmiller and Dougal[2] reported that the use of tapered inlet culverts in the state of Iowa (USA), had resulted in savings of millions of dollars in construction costs over a period of years. Specific energy curves for rectangular and circular sections were presented for the design of such structures.

The main objective of this paper is to present a case study on the improvement of an existing culvert using a tapered inlet. The effectiveness of the improved system was evaluated by hydraulic modeling.

17.2 DESCRIPTION OF THE DRAINAGE SYSTEM

The study area is located in Normanhurst, one of the northern suburbs of Sydney. Surface runoff from the northern and eastern parts of the catchment flows into a small natural creek which has a trapezoidal cross-section with base width varying from 1.5 to 3 m. A 1500 mm diameter concrete pipe culvert drains the stormwater from the south into the creek. The combined flow then discharges through a 2.4 m-wide horseshoe-shaped brick culvert under the northern railway line. At the downstream side, water flows westward through another short length (approximately 18 m) of the creek before going through two 1500 mm concrete pipe culverts under Malsbury Road which runs parallel to the railway line. The outflow then continues as free overfall into the creek.

The drainage capacity of the brick culvert under the railway line was found to be inadequate when severe flooding occurred in April 1988. It was believed that the flooding was mainly due to poor hydraulic performance, but also partly due to blockage of the culvert by debris. The flow from the 1500-mm pipe bent sharply at right angle to the railway line culvert while the flow in the creek curved around at an angle of 37°. The two inflows then merged and formed a pondage near the entrance to the brick culvert. These problems can be overcome by smoothly streamlining the inflows.

The improvement strategy involved extending the railway line culvert and bending it in alignment with the creek. This extension was designed as a 2.4 m x 1.8 m box culvert, splayed to a radius of 7.5 m. A 2.44-m long transition was needed at the connection to change the cross-section from horseshoe to rectangular shape. A tapered inlet, with a cross-section of 4.4 m x 2.32 m, was added at the entrance to streamline the flow. The 1500-mm concrete pipe was also extended and angled at 45° to join with the transition as a Y-junction. Diagram of the proposed improvement is depicted in Figure 1.

17.3 HYDRAULIC MODEL

The proposed improvement of the railway line culvert was investigated using physical modeling at the Hydraulics Laboratory, University of Technology, Sydney. The main objectives of the model study were to confirm the effectiveness of the proposed improvement to handle the design flow rate and to investigate the effect of outlet control on the water level in the creek at the upstream side of the railway line culvert. Results were reported by Patarapanich.[3]

The hydraulic model was based on the Froude number criterion for dynamic similarity, with an undistorted model to prototype scale ratio of 1:25. The corresponding ratios for velocity and discharge were thus 1:5 and 1:3125 respectively. Models of the brick culvert, concrete pipe, the transition and the tapered inlet were built with perspex. The natural conditions of the creek were

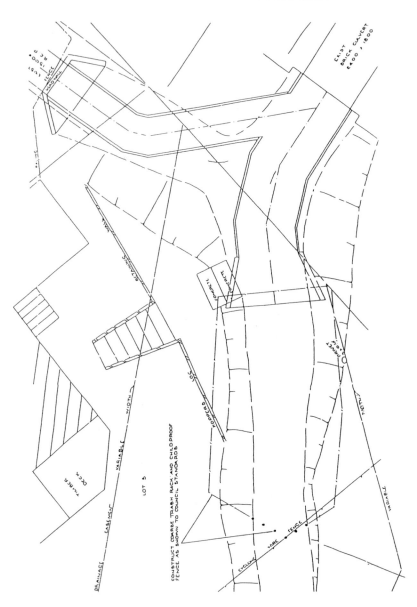

Figure 1.   Proposed improvement of the culvert.

modeled with cement mortar in a wooden box. The downstream end of the railway line culvert was connected to an outlet box from which the outflow could be adjusted with a sluice gate. The tailwater level could thus be varied to investigate its effect on the water level in the creek at the upstream side. Inflows into the creek and the concrete pipe were measured by using rotameters. Flow

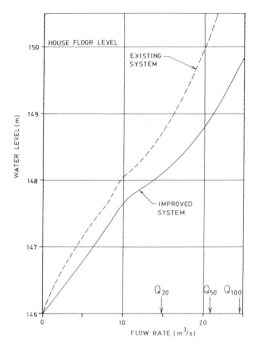

**Figure 2.   Head-discharge relationships for the existing and improved
systems.**

depth at the tapered inlet was measured with piezometer attached to the creek
bed.

17.4 DISCUSSION OF RESULTS

Catchment analysis indicated that the proportions of flows through the creek
and pipe culvert were approximately 45% and 55% of the total flow through the
railway line culvert. These proportions were used in the model test. Model of the
existing system was tested under free-fall conditions at the outlet and was
compared with the system with the proposed improvement. Variation of the
water level in the creek with flow rate is shown in Figure 2. Taking the floor level
of the house near the inlet (150 m) as the limit, the proposed improvement could
increase the capacity of the culvert from 19.8 to 25.8 m^3/sec. This represents an
improvement in flow rate of about 30%. In terms of flood level tabulated in Table
1, the proposed improvement could lower the water level in the creek by 0.7 m,
1.3 m, and 1.6 m for the 20-year, 50-year, and 100-year floods, respectively.

During the tests of the culvert with free overfall, it was observed that partially
blocking the downstream end of the culvert would cause backing up of the water
level in the creek at the upstream side. This choking of flow could actually occur

Table 1. Comparison of Flood Levels for Various Test Conditions

	Flood Recurrence Interval ARI		
	20-year	50-year	100-year
Peak discharge (m^3/sec)	14.9	20.8	24.5
Tailwater level at Malsbury Road (m)	147.25	147.5	147.6
	Water level in creek at inlet (m)		
Existing system with free overfall	148.8	150.3	151.3
Improved system with free overfall	148.1	149.0	149.7
Existing system with outlet control	149.2	151.0	152.1
Improved system with outlet control	148.5	149.7	150.5

in the prototype if the outflow was restricted by the capacities of the two pipe culverts under Malsbury Road. Preliminary calculation indicated that Malsbury Road would be overtopped by the 20-year flood, (Q = 14.9 m^3/sec) and the corresponding tailwater level (147.25 m) would be well above the invert level (145.77 m) of the brick culvert. The tailwater levels for the 50-year and 100-year floods, (147.5 m and 147.6 m) would be up to the soffit level (147.57 m) of the brick culvert. The flow through the brick culvert would therefore be controlled by the outlet condition.

To investigate the effect of the outlet control, the model was tested with the 20-year, 50-year, and 100-year floods, (Q = 14.9, 20.8, and 24.5 m^3/sec respectively). For each of these flow rates, the tailwater level was progressively raised by restricting the outflow from the outlet box. The corresponding increase of water level in the creek was noted and plotted in Figure 3. Comparison of flood levels for the various cases investigated is summarised in Table 1. The effect of the outlet control was to increase the flood level on the upstream side of the creek by 0.4 m, 0.7 m, and 0.8 m for the 20-year, 50-year, and 100-year flows, respectively. The same increases were observed for both the existing and the improved systems. Allowing for 0.3 m of free-board, the house near the inlet would be under threat by the 50-year flood. Similar results for the existing system indicated that the occurrence interval of flooding of the house was around 30 years.

17.5 CONCLUSIONS

The proposed improvement of the railway line culvert, operating under free overfall condition at the outlet, could increase the flow capacity from 19.8 to 25.8 m^3/sec when the water level in the creek was up to the house floor level of 150 m.

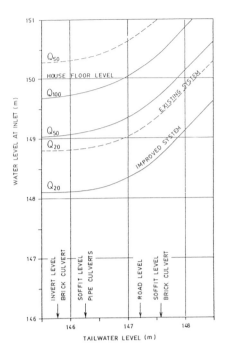

Figure 3.   Effect of outlet control on the upstream water level at the inlet.

This represented an increase in flow rate of about 30% over the existing system. In terms of water level, the proposed improvement could lower the flood level by 0.7 m, 1.3 m, and 1.6 m for the 20-year, 50-year, and 100-year floods, respectively.

The actual flow in the prototype would most likely be controlled by the tailwater level. Under this condition, the water level in the creek at the upstream side would be raised by 0.4, 0.7, and 0.8 m for the 20-year, 50-year and 100-year floods, respectively. Even with the outlet control, the proposed improvement of the railway line culvert could reduce the frequency of flooding from once in 30 years to once in 50 years.

REFERENCES

1. Pilgrim, D. H., Ed. *Australian Rainfall and Runoff, a Guide to Flood Estimation*, Institution of Engineers, Australia (1987).
2. Rossmiller, R. L., and M. D. Dougal. "Tapered inlet design using specific energy curves," *J. Hydraul. Div., Proc. ASCE,* 108: (HY1) 127 (1982).
3. Patarapanich, M. "Hydraulic model of a railway line culvert at Karinya Place, Normanhurst," Project Report No. 89/43/014, Insearch Ltd., University of Technology, Sydney (December 1989).

18 HYDROLOGIC ANALYSIS FOR THE BUKIT TIMAH FLOOD ALLEVIATION SCHEME, SINGAPORE

18.1 INTRODUCTION

The Bukit Timah Canal is a major watercourse in Singapore, originating near the highest point on Singapore Island, and flowing, via the Rochor Canal, through low-lying central business areas of the city (Figure 1). Increasing development in the upper parts of the catchment, together with lining and straightening of the original watercourse, led to frequent flooding in its urban areas near the coast, particularly at high tide.

The Bukit Timah basin has an area of 2620 ha, and lies within the core urban area of Singapore. The Bukit Timah stream has been developed and channelized over the years so that it is now entirely of a lined concrete section from Bukit Timah Hill, to the sea. Much of the lining was constructed when most of the drainage basin was still largely open space with some scattered residential development.

As the upstream areas were urbanized, the flooding problems along the main stream of the Bukit Timah Canal became quite severe. The long narrow orientation of the basin leads to stormwater runoff being quickly conveyed through storm sewers and gutters from the newly developing areas and discharged into the main canal, which did not have adequate capacity to handle these ever-increasing discharges. The result was flooding over large portions of the canal at very frequent intervals, especially near Newton Circus where the channel capacity decreases rapidly.

A Phase I diversion canal was built in 1971 to divert flow from the upper third of the basin through a transbasin diversion canal, to the Ulu Pandan system. This diversion, which essentially diverts all except dry-weather flow from the Upper

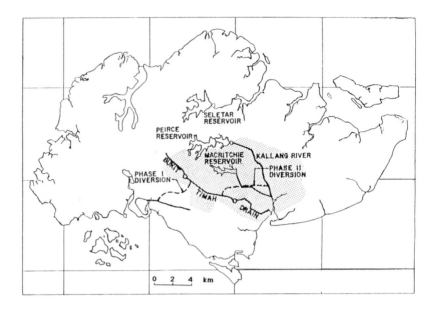

Figure 1.   Location plan in Singapore.

Bukit Timah basin, has performed very adequately since its implementation and has markedly reduced flooding downstream. However, it was recognized by the Ministry of the Environment that the Phase I diversion alone was not adequate to fully mitigate flooding along the Bukit Timah Canal.

Therefore, the Ministry of the Environment proposed the Phase II diversion canal which would divert flows from the Bukit Timah Canal at the Swiss Cottage Road to the Kallang River, and thus further reduce inflows to the critical Newton Circus area of the Bukit Timah system. Additionally, several improvements to the Bukit Timah Canal itself were proposed to increase capacity at critically undersized sections. In 1982, Camp Dresser & McKee International, Inc. was retained to conduct a computer simulation hydrologic analysis of the area, and to write an evaluation report.

18.2 STUDY OBJECTIVES

The basic objective was to develop a combined drainage system which can handle the five-year storm under future land-use conditions, and with the storm peak in the lower basin coincident with a normal high tide. In addition, the study had to make allowances for the removal of several existing control sections which limit tributary inflows to the main canal. The removal of these constraints significantly increases the flows in the lower Bukit Timah Canal and into the diversion canal.

The objectives of the approach that is used are to account for the important timing issues that become critical for complex drainage systems such as those in the Bukit Timah and Kallang Basins. For many of these cases, different times of concentrations for various tributary catchments can cause peak discharges to be out of phase and therefore not additive. Further timing changes are introduced where storm discharges are modified in passing through storage areas which exist upstream of constrictions. Time-varying outlet conditions due to tidal effects can cause flow capacities of the drainage facilities to vary during the course of a storm. All of these factors combine to make the reliable determination of the flow quantities to be used in sizing the drainage conveyance sections a difficult and complex procedure.

18.3 MODELS

Two related computer simulation models were selected for this project: the MIT Catchment Model for overall hydrograph generation, and the Two-Dimensional Flow Model to predict the peak stages reached during passage of a transient flood wave.

The MIT Catchment Model (MITCAT) is a surface runoff model that computes the runoff hydrograph from any applied rainfall event using a generalized representation of the topographic, surface cover, and infiltration characteristics of the basin involved.

The MITCAT modeling scheme is based on the philosophy that a highly detailed representation of the movement of water over the very irregular land surface and along complex channelization is neither possible nor necessary. The natural complexities of a catchment have been reduced to a number of simple elements, such as:

1. an overland flow plane, which receives a spatially uniform rainfall distribution and/or overland flow from adjacent elements
2. a stream element, taken to be a channel of uniform parameters, receiving both lateral inflow from adjacent overland flow elements and upstream flow
3. a reservoir element, used to account for storage and/or attenuation of storm runoff or discharge by natural or artificial control structures
4. a pipe element, to model storm drains or sewers in urban areas

The equations chosen to represent overland flow and stream flow were derived from the partial differential equations for unsteady flow in channels.[1,2] The resulting form of these equations is known as the Kinematic Wave Equations.

Lighthill and Whitham,[3] in their comprehensive consideration of the fluid mechanics of flood movement in rivers, separated transient effects into dynamic and kinematic waves, both of which are initially present. They showed that for Froude numbers less than two, the dynamic component decays exponentially with time, and the kinematic wave ultimately predominates. The kinematic wave equations, for overland flow or for flow in a channel, may be presented as:

1. Discharge is proportional to the cross-sectional flow area, raised to a power $[Q = aA^b]$.
2. The rate of change of discharge with distance equals the rate of lateral flow with respect to distance, less the time rate of change of cross-sectional area (which is storage per unit distance).

The estimation of the parameters a and b is performed internally in the model from readily derived properties of each flow element. These properties generally include the element's cross-sectional shape, its slope, and a roughness factor, generally the Manning n.

Use of the kinematic form of the unsteady flow equations allows particularly simple numerical solutions (since disturbances can propagate only in the downstream direction), while retaining some of the nonlinear effects of the full dynamic form. The input parameters are directly related to the physical characteristics of the watershed and, insofar as possible, are directly measurable from map or field data.

The model minimizes the amount of field data required to obtain reasonable results; the results are not overly sensitive to uncertainty in the parameter estimates. The "fudge" or "calibration" factors found in many other hydrologic models are avoided. Only a minimum of historical rainfall or runoff data are needed for calibration or estimation of parameters.

Numerical approximations such as finite grid mesh sizes, optimum time step, and linearization procedures are controlled internally to make the model as "self-sufficient " as possible, and to make it useful to engineers who are not familiar with numerical analysis or solutions of differential equations. The model can handle watersheds of any size or shape, urban or rural, as well as a wide range of hydraulic and hydrologic conditions.

MITCAT models were built for the various subbasins along the Bukit Timah and the Kallang River, so that the runoff from the total basin area from the Bukit Timah hill to the sea could be simulated. A total of 108 model segments were used to represent the entire catchment, as shown in Figure 2.

While MITCAT adequately predicts the overall flood hydrographs for the catchment, it may lack detailed accuracy in predicting flows, and particularly flood stages, in critical parts of the study area. Therefore, the MITCAT discharge hydrographs were routed through the channel system using the *Two-Dimensional Flow Model* (FLOW2D). This model provides an effective method of computing flood discharges and flood stages at various locations within a complex drainage network composed of channel reaches, constricted openings at bridges, and related flood storage regions, by simulating the movement of flood flows through the system as influenced by external boundary conditions such as tidal variations.

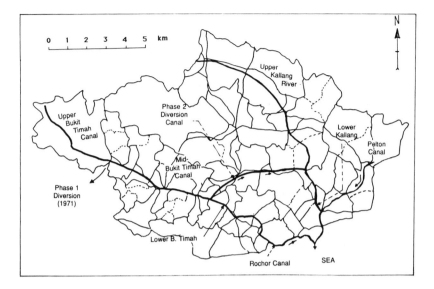

Figure 2. HMITCAT segmentation (108 segments).

It is important to note that the output from this model is the "stage" and not only the discharge values of the flow. An extremely flat channel system such as the lower Bukit Timah is largely constrained by the tidal downstream boundary conditions. The primary factor to consider during the design development is the level that the water will reach during the transient event and not only the steady-state discharge capacity of the channel. *There is little correlation between the peak stages reached and the discharge hydraulics in channels such as the Bukit Timah.*

The FLOW2D Model numerically solves the unsteady equations of continuity and momentum in finite element form to predict stages and discharges throughout the area of interest. The continuity equation relates the rate of change in water depth to the net rate of flow into an element. A simplified form of the momentum equation (Manning's formula) is used to express the balance between gravitational and frictional forces.

The physical characteristics of an estuary or channel system are defined by *segments*, *cross sections*, and *boundary conditions*, elements that are used by the model in defining the relevant hydraulic relationships involved. *Segments* are discrete areas of the study region for which a surface area vs. elevation relationship is defined for use in the continuity equation. *Cross sections* form the boundaries between adjacent segments. Hydraulic characteristics such as flow area vs. depth, flow length, and surface roughness are specified for each cross section. *Boundary conditions* are time histories of inputs to the model that drive the flows through the system. Boundary conditions may be discharge hydrographs representing storm runoff rates into the flood plain region, or water level hydrographs representing an externally controlled boundary such as a tidal variation.

On applying the model, limits of the study region are first established from the flood plain topography such that the exterior boundary is at or beyond the region susceptible to flooding. Estimated maximum flood elevations under the most severe conditions to be considered are used for this purpose. The flood plain is then divided into the segments, with hydraulic conditions specified at the crosssections.

The continuity equation is applied at each time step at each segment to relate net change in storage to net rate of inflow to the segment. The Manning formula is applied at each time step at each crosssection. The equations are expressed in nonlinear difference form; then the nonlinear expressions for discharge, Q, are linearized. The resulting set of linear simultaneous equations is solved for water surface elevations in each segment, at each time step, using a Gauss-Jordan elimination procedure.

An important limitation of this model is that N simultaneous equations must be solved if there are N segments in the study area. The practical impact of this limitation is partly offset by the capability to have irregularly shaped segments, and segments of unequal size. Values of N as large as 100 are practical to use.

A computational policy internally varies the time step size — from the order of seconds for rapidly changing water levels, to the order of hours for, say, a slowly receding flood hydrograph — to achieve desired accuracy without excessive cost in computational effort.

The main stem of the Bukit Timah Canal from the Phase I diversion downstream, the Kallang from Upper Thomson Road downstream, the diversion canal, and several major tributaries of the Bukit Timah Canal were modeled using the FLOW2D model. Since the routing of floods through these channels is highly interconnected, a single overall model was developed for this area. Various subsets of the model were used to evaluate specific local design problems. A total of 46 reaches are represented in the overall FLOW2D model, as shown in Figure 3.

18.4 SYSTEM EVALUATION

The application of the computer models to analyze the hydrologic and hydraulic characteristics of the basin yielded the following general conclusions relative to the present and required capacities of the drainage system.

18.4.1 Mid Bukit Timah Canal

The portion of the Bukit Timah Canal from Coronation Road to Swiss Cottage School is generally undersized. The design storm under present channel conditions would develop a peak discharge of about 65 m³/sec through this section of

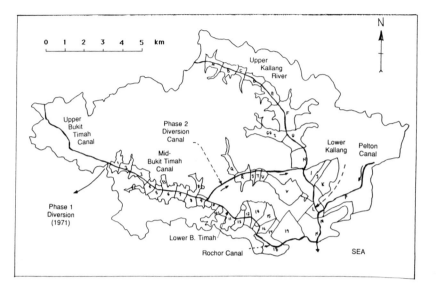

Figure 3.   Segments for FLOW2D model.

canal. Improvements will roughly double flood conveyance capacity, from 65 to 120 m³/sec.

18.4.2 Lower Bukit Timah/Rochor Canal

The canal from Newton Circus to Serangoon Road is greatly undersized, with a capacity of only 20 m³/sec in one reach. This entire portion of the Bukit Timah Canal has far less capacity than the 160 m³/sec that is estimated to be required to alleviate upstream flooding without a Phase II diversion canal.

18.4.3 Phase II Diversion Canal

The analysis indicates that flood stages in the upper portions of the diversion canal, and in the Bukit Timah Canal above the diversion, are not very sensitive to diversion canal size alone because of present outlet conditions at its junction with the Kallang River. The present invert profile of the Kallang is very restrictive in both setting the invert grades of the diversion and in developing a high tidal backwater condition throughout the length of the diversion. Under these conditions it would take large changes in the width of the diversion canal to achieve relatively small improvements in upstream flood elevations. The result would be a 26 to 35 m (depending on lower Bukit Timah improvements) diversion canal width requirement with an allowable flow depth of only slightly more than 3 m.

One effective measure to improve upon this situation would be to lower the diversion canal invert, thereby achieving the required section flow area with a smaller canal width. It is estimated, for the total diversion case, that the canal width could be reduced from 35 to 26 m by lowering the diversion invert by 1.5 m at its junction with the Kallang River. This would require dredging of the Kallang to lower its invert at this point, and for some distance downstream.

18.4.4 Lower Kallang River

The Kallang River channel below the Phase II diversion canal outlet would require several improvements to accommodate the implementation of the proposed diversion scheme. The currently unimproved reach from Jalan Toa Payoh to the Serangoon Road bridge requires widening to be able to adequately handle the additional flows from the diversion as well as the flows for the improved upstream reaches of the Kallang.

It is also essential that the Kallang be dredged so that its invert at the diversion outlet point can be lowered by at least 1.5 m. This will have the benefit of increasing the Kallang River flow section and capacity, and will lower the predicted flood profile in this section of the Kallang. The result would be to lower the peak downstream water levels at the diversion canal outlet and thereby increase its capacity.

18.4.5 Effect of the Tide

The analytical results clearly show that the effectiveness of the diversion approach is limited by high tides that exert a strong influence at the outlet of the diversion canal into the Kallang. At high rates of runoff these effects are propagated throughout the diversion canal and into the Bukit Timah Canal. Furthermore, high tide levels produce flooding in low areas of the Lower Bukit Timah region at Farrer Park and on either side of the Rochor Canal at Jalan Besar that cannot be eliminated by canal improvements alone.

18.5 ALTERNATIVES EVALUATED

Four flood-relief alternatives were evaluated, each involving various combinations of capacity increase in the various component parts of the system. Each entailed considerable capital expenditure in channel widening and in construction of the diversion canal, and each had to deal with the problem of providing flood relief at the minimal flow gradient available when design storm runoff is coincident with high tide.

The first three alternatives were all gravity flow schemes. The fourth was radically different, in that it featured a tidal barrier to be constructed at the common mouth of the Kallang River and the Rochor Canal. This facility, consisting of a pump station, a gated spillway, and a navigation lock, would be operated to maintain a fairly constant pool level in the Kallang River and Rochor Canal, and isolate the system from the influence of high tides. The lower tailwater conditions would provide additional hydraulic gradient to increase the flow capacity of the upstream canals.

Overall costs for the four alternatives were similar, with a range of about 15% between the lowest cost scheme and the most expensive one. The pumping scheme, alternative 4, was lower in initial construction cost, but higher in its estimated operations and maintenance costs.

18.6 CONCLUSIONS AND RECOMMENDATIONS

The models indicated the extent to which upper reaches should be widened; the gross inadequate capacity of the lower reaches; and the fact that the Phase II diversion canal, built in a low, flat area, would either need to have a very large cross section, or be provided a stormwater pump station, to adequately pass flood flows occurring at high tide.

Use of such models was critical to the study since the primary flood-producing events are short and intense, leading to highly transient varying flows in the canal. In many cases, particularly at high tide, the lower reaches of the canal act as a large reservoir, and never develop hydraulic conditions which approximate steady-state flow. The models showed that the transient response characteristics are such that more cost-effective solutions can be developed than those that would be indicated by traditional, steady-state approaches.

The Phase II diversion canal, not yet with a pump station, has been completed, and was officially opened in August 1990. To date, it has successfully met all expectations.

The feasibility of a pumping barrage for tidal protection of the Bukit Timah drainage system and for other nearby streams has been studied further. A decision to implement is pending.

REFERENCES

1. Eagleson, P. *Dynamic Hydrology* (New York: McGraw-Hill, 1970).
2. Harley, B.M. "A Modular Distributed Model of Catchment Dynamics," Massachusetts Institute of Technology, Parsons Laboratory, Report No. 133 (1970).
3. Lighthill, M.J., and G.B. Whitham. "On Kinematic Waves-I. Flood Movement in Long Rivers," Proceedings of the Royal Society of London (1955), pp. 229, 281.

19 MANAGEMENT OF URBAN STORMWATER COLLECTION SYSTEMS — AN OVERVIEW

19.1 INTRODUCTION

The commissioning of the Sungei Seletar-Bedok Water Scheme in 1986, at a cost of S$277 million,* was a new milestone in the Public Utilities Board's continuous efforts to meet the increasing demand for water. The scheme has been developed to exploit the hitherto untapped catchments of the north-eastern and eastern parts of Singapore. A unique feature which sets this scheme apart from the Board's earlier schemes is the utilization of storm runoff from urbanized areas as a source of raw water.

The main components of the scheme are as follows (see location map in Figure 1):

1. Sungei Seletar Reservoir
2. Bedok Reservoir
3. Stormwater Collection Systems
4. Bedok Waterworks

The Sungei Seletar Reservoir is located in the northeastern part of Singapore. It was formed by the construction of an earthen embankment dam across the mouth of the Sungei Seletar. In contrast, the Bedok Reservoir, located north of Bedok New Town, was developed from a former sand pit. These two reservoirs, besides catching runoffs from their natural catchments, also collect surface runoff from the urbanized areas of Ang Mo Kio, Bedok, Tampines and Yishun New Towns, and the northwestern part of Changi International Airport. This is done through a complex collection and conveyance system comprising nine collection ponds, nine pumping stations, connecting pipelines, and a gravity flow diversion channel.

---

* 1 U.S.$ = 1.7 S$ (approx.)

0-87371-805-4/93/$0.00+$.50
© 1993 by Lewis Publishers, Inc.

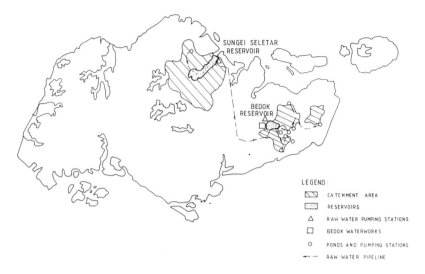

Figure 1.   Location map of Sungei Seletar and Bedok reservoirs/storm water collection facilities.

Since then, other stormwater collection systems have also been constructed, or are being planned, in other parts of the island, namely in Bukit Panjang, Woodlands, and Hougang New Towns. While earlier stormwater collection systems implemented under the Sungei Seletar-Bedok Water Scheme were designed mainly as utility ponds, the newer systems have incorporated improved aesthetics, landscaping, and some recreational facilities into their design.

This paper discusses the various aspects and requirements in the planning, implementation, operation, and maintenance of urban stormwater collection systems in Singapore.

19.2 SYSTEM CONCEPTS

The primary objective of a stormwater collection system is to maximize surface runoff collection. Parameters that have significant impact on such collection are rainfall patterns, catchment/drainage layout, and land-use characteristics.

Rainfall in Singapore is characterized by relatively high-intensity and short-duration storm events of small areal distribution. Such storms falling on small catchments will generate highly peaked and sudden storm flows, normally termed as flash flows. Such storm flows will have very short times of concentration and will leave the basin rapidly. With a higher percentage of impervious covers through urbanization, improved storm drains and lined drainage channels, the storm runoffs will leave the basin even faster.

Urban stormwater collection systems must therefore be designed to catch the "flashy" runoffs in a relatively short period. In this connection, the more established direct abstraction stations would not be efficient. Modifications/additional features would be required, namely:

1. the need for a holding pond to temporarily store the storm runoff before its transfer to the main impounding reservoir(s)
2. the need for a diversion structure to channel the storm flows from the storm drains to a temporary holding pond

Through such an arrangement, an estimated capture of up to 70% of the available runoff is possible.

19.3 DEVELOPMENT

Stormwater, if it is to be used for water supply augmentation, must be of acceptably good quality. Undesirable discharges, such as from industries and other pollutive activities, must therefore be excluded. This could be realized through:

1. judicious land-use planning and its control within the catchments to be developed as stormwater catchments
2. incorporation of designs/measures in town planning to minimize pollution
3. the enforcement of pollution control measures

Toward this end, an integrated and coordinated approach to urban town planning has been adopted through close liaisons with the various relevant developmental bodies involved, such as the Planning Department, the Housing and Development Board (HDB), the Ministry of the Environment (ENV), and lately, the Parks and Recreation Department (PRD).

Notable features of such coordinated planning at the macrolevel are as follows:

1. careful planning of land use and its control to exclude pollutive activities, principally industrial, from the proposed stormwater catchments
2. construction of covered drains and gutters around HDB blocks to prevent entry of litter/debris into the drainage system
3. redesign of HDB bin collection centers and refuse chutes to minimize spillage
4. grading of HDB void decks for discharge into public sewers
5. strict enforcement of antipollution laws and regulations
6. provision for greater green spaces and/or recreational areas

At the microlevel, close liaisons are maintained to ensure that the main components of the collection system, such as diversion structures and holding ponds, are incorporated into the main drainage network. In addition, the collection system itself, principally the combination of diversion structures, pond sizes, and pumping facilities, is designed for optimal operation via computer mathematical modeling techniques. Up to 70% of the storm runoff can

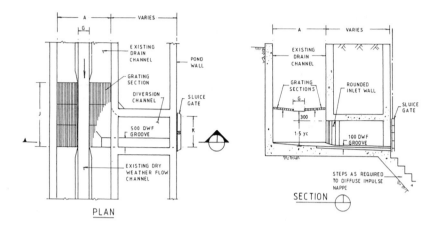

Figure 2.   Drop inlet structure.

be collected via the optimized design of the diversion structures, ponds, and pumps for each of the stormwater catchments.

Beyond the utilitarian/functional design of the initial group of stormwater collection systems, the newer ones being constructed/planned have been upgraded to incorporate aesthetics, landscaping, and even limited recreational facilities into their design. These systems would now form part of a town garden or park. Besides providing temporary storages, they would also serve as scenic water bodies with recreational/amenity functions.

19.4 OPERATION OF STORMWATER COLLECTION SYSTEMS

Additional factors to consider in the development and management of urban stormwater collection systems, besides water quality aspects, would include both operations and maintenance aspects of these systems. The principal components are the diversion channel, its related equipment, and the pumping and ponding facilities.

As an illustration, the two main types of diversion structures as installed under the Sungei Seletar-Bedok Water Scheme are drop inlets (Figure 2) and controlled barrage gate regulators (Figure 3).

In design, both types of structures incorporate a bypass flow channel which prevents the lower quality dry-weather flow and the first part of any storm flow, called the "first flush", from entering the diversion channel. Accumulated pollutants carried by the "first flush" from the catchments are therefore not collected.

From the operational viewpoint, drop inlets are preferred. They are easier to operate and maintain as storm flows are collected in a "passive" mode and no

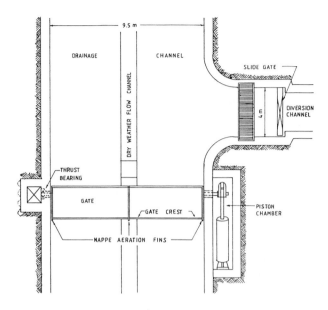

Figure 3.   Barrage gate structure.

mechanical operations are required. In contrast, collection through barrage regulators involves mechanical operations. Typically, as shown schematically in Figure 4, the barrage would consist of a steel gate, hinged at the bottom. The barrage is left down in the "open" position in dry weather, but is raised (closed) to divert flows during normal design storm events. However, the barrage gate can be lowered rapidly as required, to enable flood flows in excess of the diversion capacity to pass. In terms of operation, during a storm event, a level element in the main drain senses the rising flow level and initiates closure of the barrage gate. An additional level element installed in the diversion channel downstream of the sluice gate allows for comparison of the two levels. If the drain level is more than 50 mm above the channel water level, the sluice gate will open to divert flow into the reservoir. For structures located within tidal influence, e.g., at Yan Kit (see Figure 5), each of the structures is further equipped with a gate on the dry-weather flow channel to prevent backflow of low quality, and possibly saline, water into the pond. In addition, a conductivity probe is installed in the dry-weather flow channel just upstream of the sluice gate to monitor the salinity levels.

In consideration of the tight labor situation, operation of stormwater collection systems in Singapore is highly automated. The collection systems are designed, as far as possible, for unmanned operations. Remote monitoring and remote control of these systems are possible via computer-based Supervisory Control And Data Acquisition (SCADA) systems. In this respect, a centralized control center, equipped with back-to-back computers and associated telemetric periph-

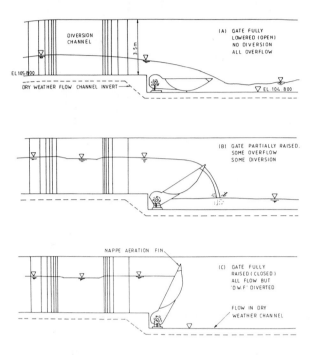

Figure 4. Lateral diversion – views of barrage gate in several positions.

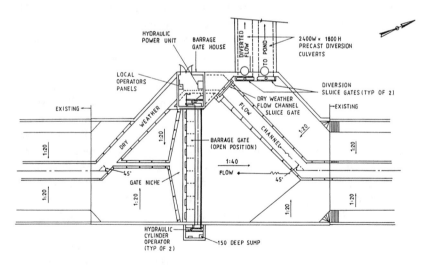

Figure 5. Yan Kit diversion.

erals is installed at Bedok Waterworks to continuously monitor parameters, such as water levels in the collection ponds and diversion channels, pumped flow rates to the impounding reservoirs, status of pumps, and positions of barrage gates. The Center also enables the various plants and equipment to be remotely controlled.

19.5 WATER CATCHMENT MANAGEMENT PROGRAM AND MAINTENANCE OF STORMWATER COLLECTION FACILITIES

In general, water catchment management programs are designed to ensure that good, acceptable quality water is collected.

For urban catchments, the main activities likely to contribute to water pollution, and hence adverse impact on water quality, are:

1. illegal dumping of toxic wastes, oil, and chemicals
2. poor sullage and sewerage systems at construction sites, as well as remnant squatter houses
3. pollution loads from vehicular movements and accidents that cause oil/ chemical spills

As such, the above activities are tightly controlled through strict legislation and regulations, such as the Water Pollution Control and Drainage Act, the Sanitary Plumbing and Drainage Systems Regulations (1976), the Surface Water Drainage Regulations (1976), the Environmental Public Health Act (1975), the Trade Effluent Regulations (1976), the Toxic Industrial Wastes Regulations (1988), the Public Utilities (PU) Act Chapter 211, together with the PU (Water Supply) Regulations (1977), and the PU (Catchment Area Parks) Regulations (1989). Towards this end, close liaisons with other authorities, e.g., the Planning Department and the HDB and ENV, are maintained.

Maintenance aspects, such as desilting of stormwater collection ponds and servicing of electrical and mechanical plants and instruments, play an important role in the maximization of the collection of stormwater from urban catchments. In catchments where there are many developments, the storm flows contain much higher silt loads, thereby causing rapid siltation. Regular desilting of all stormwater collection ponds and diversion channels, at least once a month, is required. This is carried out by contract labor. The main drains, together with the drop inlets, would have to be inspected once every two days to ensure that all accumulated debris are removed. This is done to prevent chokage.

To ensure proper, consistent, and efficient performance of the electrical and mechanical plants, a comprehensive preventive maintenance program is in place. Operating parameters of pumpsets in the pumping stations such as oil level of the oil lubricated bearings, bearing temperature, gland packing seepage, excessive noise, or vibration are checked daily. In addition, the pumps and motors are also checked monthly, for misalignment, abrasion of slip-ring, and the cleaning of parts soiled with carbon rust. For the smooth operation of the barrage gates, routine inspection, including exercising of the moving parts and hydraulic system, is carried out on a quarterly basis. All servicing, such as cleaning and greasing of wire ropes and remedial works, e.g., paint work of the hoists, replacement of broken element wires, tightening of loosened nuts and bolts, etc., are also carried out.

For the level monitoring systems, including the conductivity probes, both preventive and corrective maintenance are carried out. Preventive maintenance would include monthly inspection and maintenance of the condition of the primary sensor, signal cables, and the zero and span settings of the level converter. Corrective maintenance usually involves attending to the common problems of blown fuses in the converter, the drifting of the zero and span settings of the converter, and choked or punctured primary sensor(s).

19.6 CONCLUSION

Urban stormwater collection systems developed under the Sungei Seletar-Bedok Water Scheme, play an important role in augmenting Singapore's water supply. Through integrated and judicious planning, close liaisons with other development authorities, and the close monitoring and control of pollution, relatively good raw water quality, acceptable for its intended use, can be maintained. Beyond the functional/utility design of earlier stormwater systems as developed under the Sungei Seletar-Bedok Water Scheme, the newer collection systems being planned or under construction in Bukit Panjang, Woodlands, and Hougang New Towns will have improved/additional features, such as better aesthetics and landscaping, as well as some recreational facilities, incorporated into their designs.

STORMWATER RECLAMATION

20 RECLAMATION OF URBAN STORMWATER FOR AQUACULTURE

20.1 INTRODUCTION

The City State of Singapore has a total land area of around 620 km^2 of which 52% are urbanized and 48% reserved for water catchment and park and recreation. Only pockets of small strips of land remain for agriculture and fish and prawns farming. The climate is warm (26 to 32°C), humid (80%), and equable. There are two dominant monsoon seasons: the SW monsoon from April to October, and the NE monsoon from November to March. The long-term climatological data based on records kept by the Singapore Meteorological Service are given in Table 1. Although the precipitation rate is high, water loss through evapotranspiration is equally high. Water conservation is essential to meet water needs for domestic, industrial, and agricultural activities.

The agro-industrial water requirement is typically high. For freshwater fish cultivation, it is important to have reliable sources of good quality water. The development of improved stormwater collection and management practices is a major challenge for any fish farming activities in Singapore. This paper describes the experience of water management practices of a 2-ha experimental fish farm located at the junction of Ponggol Road and Fishing Port Road between the Ponggol and Serangoon Rivers. The farm was originally developed for pig rearing. Pig farming was phased out in November 1989. The facility is being slowly cleared and converted for the cultivation of freshwater fish. The total drainage area for stormwater collection is around 20 ha. Rainwater is collected through a network of open drains leading to an earthen pond. As the experimental farm is located near the Johor Strait, the land is flat and low lying. All stormwater flow is gravitational and energy for pumping is not necessary. When water is tapped to fill the fish ponds, there is adequate head (>2 m) for filtering the water through a filtration bed and subsequent filling of the storage pond.

0-87371-805-4/93/$0.00+$.50
© 1993 by Lewis Publishers, Inc.

Table 1. Climatological Data

Parameters	Day	Month	Year
Temperature, °C			
Daily mean	—	25.6–27.4	26.6
Mean daily max.	—	29.8–31.3	30.7
Mean daily min.	—	22.9–24.4	23.8
Extreme max.	34.8	33.4–34.8	—
Extreme min.	19.6	19.6–21.7	—
Precipitation, mm			
Average	—	161–280	2402
Highest	432	379–819	3452
Lowest	—	8–92	563
Rain days			
Average	—	13–19	180
Maximum	—	20–26	222
Minimum	—	2–13	142
Evaporation, mm			
Mean	—	110–163	1659
Highest	—	135–234	1894
Lowest	—	81–123	1407
Solar radiation, $mW \cdot h/m^2$			
Daily mean	—	0.039–0.052	0.046
Highest month	—	1.31–1.82	—
Lowest month	—	1.04–1.40	—

20.2 MATERIALS AND METHODS

The farm was built by converting the wastewater treatment facilities of a pig farm having a standing pig population of 35,000. All stormwater collected through a network of storm drains is filtered with a filtration bed packed with sand and nylon mats before it is stored at the water storage lagoon which has an effective volume of 6600 m^3. Maximum depth of the lagoon is 5 m. All freshwater needs of the farm are drawn from this lagoon. The anaerobic lagoon for pig wastewater treatment was cleaned and disinfected before it was used for fish rearing. This lagoon has a concrete-lined side slope of 1:3, a maximum depth of 5 m and an effective volume of 25,000 m^3. In addition to floating fish-rearing cages in the center of the lagoon, rearing cages were also built on the sides. The sludge drying beds of the original pig waste treatment facilities were converted to five fish ponds each having an effective volume of 1200 m^3 (600 m^2 surface area and 2 m average depth). To reduce the water demand from the stormwater storage pond, water is recirculated between these fish ponds and the large fish-rearing lagoon through a system of submersible pumps and pipe work. This system also allows the discharge of accumulated sludge at the ponds and lagoon at regular intervals. Discharges of bottom sludges are carried out once every other day at a rate equivalent to approximately 5% of the pond water volume. Sludge accumulation at the large lagoon is slow and desludging will take place probably once in three years. The ponds are currently stocked with an average of 6000 fish per pond. The average size of each fish is 100 to 300 mg. The marketing size of the fish is to be around 800 mg. Density of fish near marketing size will be trimmed to 3000 to 4000 per pond by transferring the fish at stages to the cages at the large lagoon.To maintain the dissolved oxygen level in the water at an acceptable level floating aerators are installed at the ponds and the lagoons. A network of diffused air system is also installed to supply air to the fish cages at the sides of the lagoon. Installed horsepower at the ponds is around 1 kW per pond. Four 1 kW air compressors supply the needed air to the side cages at the lagoon. Two 1 kW aerators are placed near the floating cages at the centre of the lagoon. No treatment facilities are installed at present for the water in the lagoon and the ponds.

All physical, chemical, and biological parameters are determined in accordance with the *Standard Methods*.[1]

20.3 RESULTS AND DISCUSSIONS

The conversion of the pig waste-treatment facility to fish-rearing ponds took around 7 months. Clearing the accumulated sludge in the lagoon and subsequent disinfection alone took up to 4 months. Complete filling of the water storage pond, the fish rearing lagoon, and the ponds was carried out in stages over a 4 to 6 month period. Including the cleaning and flushing of lagoons, ponds, and plant facilities, around 48,000 m^3 of stormwater were collected during this period. Water quality from the storm drain was relatively good (Table 2). The pH was between 6.5 to 8.1, with most of the days measured at around 7.2

There were no detectable cadmium and lead levels, based on analysis using the atomic absorption spectrophotometer with a graphite furnace. Other heavy metals, such as aluminum, chromium, copper, manganese, nickel, and zinc, were all less than 100 ppb. Ammonia, nitrite, and nitrate nitrogens, were in general, less than 5 mg/L. Chloride averaged around 70 mg/L, and sulphate 85 mg/L. Five-day biochemical oxygen demand (BOD) was less that 3 mg/L, and chemical oxygen demand (COD) less than 78 mg/L.

There was little improvement in water quality after it was passed through the filter bed. Nitrogens and phosphate content, however, appeared to have decreased. Water quality at the storage lagoon did not change significantly from that of the filter effluent during the period studied. The water at the storage lagoon serves as the main source of fresh-water supply for replenishing the fish-rearing lagoon and the fish ponds. Overflow from the storage lagoon discharged directly into the fish-rearing lagoon. Recycling of water was carried out between the fish ponds

Table 2. Water Quality

Parameters	Drain	Filter Effluent	Tilapia Pond	Sea Bass Pond	California Trout Pond	Fish Lagoon
pH	6.5–8.1	7.2–9.4	7.0–9.4	7.2–8.2	7.6–8.8	6.9–9.6
BOD mg/L	0.6–2.8	0.6–4.8	ND–6.7	2.0–7.0	2.0–7.0	0.1–1.9
COD mg/L	27–78	27–121	41–139	38–126	45–135	17–64
TOC mg/L	8–17	9–18	9–18	11–18	7–8	8–9
SS mg/L	2–15	5–87	6–204	20–336	10–83	3–42
VSS mg/L	ND–11	1–79	6–204	16–228	3–100	1–13
PO_4 mg/L	ND–22	ND–1.4	ND–6	ND–3	ND–2.1	ND–13
SO_4 mg/L	21–144	75–220	143–536	150–220	101–540	211–541
Cl mg/L	18–132	174–283	286–389	326–1245	365–1102	921–4548
Na mg/L	1–448	7–204	163–278	600–4100	340–5924	340–5924
Ca mg/L	22–100	14–66	30–159	53–157	51–153	79–625
Mg mg/L	3–69	18–52	24–244	24–244	24–244	81–292
Mn μg/L	1–75	ND–23	1–22	1–22	1–26	ND–23
Al μg/L	ND–17	ND–87	ND–20	ND–23	ND–20	ND–11
Cr μg/L	ND–7	ND–7	ND–5	ND–6	ND–3	ND–5
Cu μg/L	3–66	4–66	2–58	3–59	4–187	4–58

Fe µg/L	58–662	15–223	27–157	15–869	15–38	5–127
Ni µg/L	ND–12	ND–8	ND–16	ND–19	ND–19	ND–15
Zn µg/L	ND–64	ND–121	ND–56	7–63	4–113	9–83
Nitrogens						
NH_3 mg/L	ND–13	ND–0.3	ND–2.8	ND–3	ND–3	ND–0.8
NO_2 mg/L	ND–3.6	ND–3.9	ND	ND	ND	ND
NO_3 mg/L	ND–6.6	ND–5.5	ND–11	2–12	ND–9	0.1–3.5

and the fish-rearing lagoon. Periodically, water at the fish ponds was replenished with water from the storage lagoon. With regular desludging, the hydraulic retention time for freshwater flow through the fish ponds was estimated to be around 40 d.

Fish population at the tilapia pond, the California trout pond, and the sea bass pond were, respectively, 6000, 6300 and 3400. Daily feed of commercial fish meal to these ponds were around 60 kg (wet weight) for the tilapia and the California trout ponds, and 45 kg (wet wt) for the sea bass pond. Growth rates of these fish were surprisingly uniform at around 0.8 to 1.3 g/fish/d. The growth efficiency (total weight of the fish per total weight of the feed) is estimated to be 25% for tilapia, 26.3% for California trout, and 18.9% for sea bass. Around 30 m^3 of accumulated sludge at 1400 mg SS/L were removed per day. Suspended solids (SS) in these ponds varied from around 10 to over 300 mg/L, BOD below 7 mg/L, TOC less than 18 mg/L and COD just around 100 mg/L. At this stage of the studies, no treatment facility was installed for the recycling water. Natural assimilation at the fish cultivating lagoon and biodegradation through a controlled aeration process at the fish ponds were the main mechanisms for maintaining the water at an acceptable quality. It was observed that the water quality did not fluctuate significantly. SS concentration level was higher at the fish ponds (in general, mean SS concentration >150 mg/L) due to the growth in microorganisms, such as bacteria and algae. More than 80% of the algae were chlorella. The pH during daylight often measured greater than 8.3, a typical phenomenon of algae activity.

The ammonia and nitrate remained at relatively low concentration levels. No nitrite was detected. Similarly, phosphate concentration remained at relatively low levels. The aerators were able to maintain the DO level at around 5 mg/L down to mid-depth of the fish ponds. Even at the bottom layer DO was observed to be around 0.5 mg/L. Nitrogens and phosphates are essential for cell growth. They were taken up by the fish, and a large part of the nutrient input to the system from fish feed were also removed through sludge discharges. Thus, through a combination of controlled aeration, fish harvesting, water recycling, water replenishment, and desludging, it was possible to maintain the nutrient concentration at predetermined levels.

In spite of continuous reuse of the water, no accumulation of heavy metals was observed. Aluminum, chromium, copper, nickel, zinc, and iron concentrations remained at relatively low levels. The concentrations of these substances were, in general, within the limits of the U. K. Environmental Quality Standards for the protection of fish and invertebrates at inland and coastal waters.[2] Lead, mercury, and arsenic were not detected in the farm water. There were occasions, especially at the early stages of the farm operation, when copper and zinc concentrations were greater than 100 ppb and the hardness of the water was in the range of 150 to 200 mg/L. The bottoms of the ponds and lagoons were lined only with clay and coarse gravels (equivalent diameter of around 4 cm). These facilities had been used for pig waste treatment. In spite of the clean-up there remained some pollutants that might have leached into the water. However, little or no evidence of metal toxicity was observed. At the early stages of the farm operation, fish attrition rate was greater than 60%, mainly due to the lack of oxygen before proper aeration equipment could be installed. The California trout also suffered from *Vibrio sp.* infection that was the main cause of California trout death. The disease has since been controlled. After the installation of proper aeration equipment, survival rate of the fish has been greater than 70% until a size of around 300 mg has been reached. Chloride and sodium concentration levels at the fish lagoon and the fish ponds were high. This was the result of an attempt to keep the water in the fish lagoon at a chloride concentration level of 2000 mg/L by pumping seawater into the lagoon at the start of the farm operation. This was one method tested at laboratory studies and found to be effective in the control of certain diseases of fresh water fish. Sea bass is a saltwater variety but was successfully cultivated in a freshwater environment where the chloride concentration was kept at around 4000 mg/L at the beginning and gradually reduced to below 1000 mg/L after a period of two months.

Fish population density at the fish lagoon was still very low during the period studied. The fish lagoon behaves more as a storage reservoir, where nutrients and BOD input are low and natural assimilation capacity is able to keep the water quality at relatively good conditions. BOD was in general lower than 2 mg/L and COD averaged around 35 mg/L and TOC <10 mg/L. The pH during sunlight reaches 9.6 at times, although there were no eutrophication problems. Nitrogens were low and phosphate averaged around 3 mg/L. Heavy metals were, in general, at low concentration levels. The high chloride and sodium contents were results of direct pumping of seawater into the lagoon. The average chloride concentration at the fish lagoon was around 1000 mg/L toward the end of this study period. In spite of recirculating the fish pond water to the lagoon, water quality at the lagoon remained relatively good with little quality fluctuation.

20.4 CONCLUSION

A study was carried out to reclaim stormwater from a 20 ha site for the cultivation of tilapia, sea bass and California trout. Sea bass is a saltwater variety, but was successfully cultivated in a freshwater environment. The water collected from the storm drains met the water quality requirements for cultivating these three species of fish. With controlled fish population density, proper aeration, and regular desludging, it was possible to maintain the water quality at an acceptable level. At a density of around five fish per cubic meter (size of fish around 300 mg) survival rate of fish was above 70%. The mean temperature of the pond water was around 28°C and all three species of fish seemed able to adapt to the condition.

REFERENCES

1. *Standard Methods for The Examination of Water and Wastewater*, 17th Ed. (Washington D.C.: American Public Health Association, 1989).
2. Mance, G., and O'Donnell. "Application of the European Communities directive on dangerous substances List II substances in the UK," *Water Sci. Technol.* 16:159 (1983).

21   RECLAMATION OF
URBAN STORMWATER

21.1 INTRODUCTION

As population and industry grow, water demand increases, and water supply becomes more of a problem. It has been estimated that the total gross water intake for all purposes in the U.S. will exceed the total available water supply of 650 billion gal/d (2.5 billion m^3/d) by the year 2000. Therefore, wastewater reclamation will become a more significant means of augmenting water supply.

Reclamation of municipal wastewater for industry, nonpotable domestic usages, and groundwater recharge has been practiced in the United States over the past several decades. In a 1971 U.S. EPA nationwide survey, it was estimated that current reuse of treated municipal wastewater for industrial water supply, irrigation, and groundwater recharge was 53.5 billion gal/year, 77 billion gal/year, and 12 billion gal/year, respectively.[1] It is reasonable to expect that the reuse of the treated wastewater for industrial cooling, nonpotable domestic water supply, and park and golf course irrigation will be substantially increased in the future. Publications are available on the reuse of municipal effluent for potable and nonpotable water supply.[2-4] However, very limited information is available on reuse and reclamation of urban stormwater. This paper examines current urban stormwater control and treatment technology leading to the feasibility of urban stormwater reuse for various purposes, including industrial cooling and process, irrigation, and recreational water supply. The hydrologic and hydraulic aspects of urban stormwater systems are presented elsewhere and are not addressed here. A hypothetical case study illustrating the cost-effectiveness of reclaiming urban stormwater for complete industrial supply is presented.

21.2 QUALITY AND QUANTITY OF URBAN STORMWATER
AND COMBINED SEWER OVERFLOW

In recent years, a considerable number of characterization studies have been performed on urban stormwater. The reported stormwater quality parameters vary considerably in concentration and mass. Storm runoff volume and flow rate

0-87371-805-4/93/$0.00+$.50
© 1993 by Lewis Publishers, Inc.

vary greatly as well. These variations occur not only with time as the storm progresses, but also from location to location during the storm. Additional and significant influences on quality are attributable to drainage system configuration, antecedent dry-weather period (allowing pollutant accumulation), and degree of imperviousness. Because of these multiple variations and the difficulties associated with representative sampling, relationships between cause and effect are obscured, even though a considerable amount of data are available. Table 1 summarizes a comparison of quality of stormwater discharges for various cities with that of untreated and treated municipal sewage.[5]

The most obvious conclusion about the quality of urban stormwater is that it varies greatly from one metropolitan area to another. The data also indicate that the 5-day biochemical oxygen demand (BOD_5) concentration is very close to that of secondary treatment plant effluent, while the suspended solids concentration is similar to that of raw sanitary wastewater.

A similar comparison for combined sewer overflow (CSO) quality is presented in Table 2.[5] CSO is a mixture of various proportions of municipal sewage and storm runoff. For this reason, it would seem that the pollutant concentration of CSO would lie between that of the local sanitary sewage and urban storm runoff. The quality of sanitary sewage follows cyclical patterns over daily and weekly periods according to local demographic conditions. Therefore, the pollutant concentration of CSO may be affected significantly by the hour of storm occurrence. Marginally-sloped combined sewers encourage significant quantities of dry-weather sanitary sewage solids deposition that accumulates during the low-flow dry periods, and is subsequently flushed out by high storm flows creating shock effects.

21.3 CONTROL AND TREATMENT PROCESSES

Many unit processes designed for treatment of water and wastewater can be adopted to stormwater treatment. Because of the high volume and variability associated with storm and CSO, high-rate physical treatment units are considered to be advantageous over biological systems in many situations. Physical treatment alternatives have demonstrated the capability to handle high and variable flow rates and solids concentrations, whereas biological processes are more vulnerable to these variable flow conditions, and the relatively high concentration of nonbiodegradable solids in storm flow. Therefore, physical treatment systems are emphasized in this analysis.

Systems for urban stormwater reclamation may include storage and pretreatment, secondary treatment, and advanced treatment processes. Selection of treatment level is dependent on the reclaimed water usage. For example, water supply for irrigation requires a minimum degree of treatment, while that for boiler feed requires a high level. The various units for these basic processes are

(1) control-storage or retention basin, and swirl regulator/concentrator; (2) pretreatment-bar/coarse screen, swirl concentrator, and sedimentation; (3) secondary treatment combination of sedimentation, dissolved air flotation, or microscreening with biological treatment; and (4) advanced treatment screening, filtration, and carbon adsorption.

The following sections describe treatment processes using test data obtained from stormwater control demonstration projects supported by the EPA Office of Research and Development Storm and Combined Sewer Overflow Pollution Control Program.[5-7]

21.3.1 Storage

Because storm flow is unsteady and intermittent, storage facilities are a necessary consideration for the system. Storage facilities may be constructed in-line or off-line. The in-line storage concept is based on the use of the excess trunk or interceptor sewer capacity or other structures for temporary detention of flow. Subsequently, the flow is released by gravity (without pumping) to the sewage (or stormwater) treatment plant.

Off-line storage, which requires influent or effluent pumping, is used to attenuate storm flow peaks, and also provides some degree of treatment by enabling suspended solids to settle out. A multitude of storage facility types and configurations have been planned and constructed, including lagoons, and earth-lined and unlined basins; open and covered concrete tanks; underground silos; deep tunnels; underwater bags and reinforced plastic curtains suspended from wooden pontoons in the receiving-water body; and void space storage (with overhead land used for other purposes). Concrete tanks are the most commonly-used storage facility for CSO, whereas earthen basins are usually employed for separate stormwater runoff storage.

The city of Boston has constructed the Cottage Farm Stormwater Detention and Chlorination Station which is designed to provide a minimum 10-min detention and chlorine contract to the 1-in-5 year CSO rate of 233 mgd (883,000 m^3/d);[8] however, the facility will provide complete capture or longer detention of the majority of flow coming from the more frequent storm events. The facility consists of bar screens and coarse screens, a pumping station, a hypochlorination facility, detention basins, fine screens, and an outfall. Flow is pumped to the detention tanks, each of which is gated to allow isolation for easier washdown and maintenance. Hypochlorite solution is fed to the pump discharges. To reduce overflow solids carryover, detention tank effluent passes through fine-mesh screens (<100-mesh openings) before discharge via the outfall. Screenings and settled solids are discharged to the downstream interceptor and finally conveyed to the domestic wastewater treatment plant. Because of the success of the Cottage Farm Detention and Chlorination Station, the Charles River marginal conduit

TABLE 1. Comparison of Quality of Storm Sewer Discharges for Various Cities[a]

Type of Wastewater, Location, and Year	BOD$_5$ in mg/L		COD in mg/L		DO in mg/L	Suspended Solids in mg/L		Total Coliforms in MPN/100 mL		Total Nitrogen, as N, in mg/L	Total Phosphorus, as P, in mg/L
	Average	Range	Average	Range	Average	Average	Range	Average	Range	Average	Average
Typical Untreated Municipal	200	100–300	500	250–750	—	200	100–350	5×10^7	1×10^7–1×10^9	40	10
Typical Treated Municipal Primary Effluent	135	70–200	330	165–500	—	80	40–120	2×10^7	2×10^6–5×10^8	35	7.5
Secondary Effluent	25	15–45	55	25–80	—	15	10–30	1×10^3	1×10^2–1×10^4	30	5.0
Storm sewer discharges: Ann Arbor, MI, 1965[2]	28	11–62	—	—	—	2,080	650–11,900	—	—	3.5	1.7
Castro Valley, CA, 1971–1972[14]	14	4–37	—	—	8.4	—	—	2×10^4	4×10^3–6×10^4	1.9[b]	—
Des Moines, IA, 1969[6]	36	12–100	—	—	—	505	95–1053	—	—	2.2	0.87

Location											
Durham, NC, 1968[1]	31	2–232	224	40–660	—	—	—	3×10^5	$3 \times 10^3 - 2 \times 10^6$[c]	—	0.18
Los Angeles, CA, 1967–1968[19]	9.4	—	—	—	6.9	1013	—	—	$3 \times 10^3 - 2 \times 10^6$	—	—
Madison, WI, 1970–1971[17]	—	—	—	—	—	81	10–1,000	—	—	4.8	1.1
New Orleans, LA, 1967–1969[15]	12	—	—	—	4.5	26	—	1×10^6	$7 \times 10^3 - 7 \times 10^8$	—	—
Roanoke, VA, 1969[12]	7	—	—	—	—	30	—	—	—	—	—
Sacramento, CA, 1968–1969[37]	106	24–283	58	21–176	—	71	3–211	8×10^5	$2 \times 10^4 - 1 \times 10^7$[c]	—	—
Tulsa, OK, 1968–1969[33]	11	1–39	85	12–405	—	247	84–2,052	1×10^5	$1 \times 10^3 - 5 \times 10^8$	0.3–1.5[e]	0.2–1.2[f]
Washington, D.C., 1969[5]	19	3–90	335	29–1514	—	1697	130–11,280	6×10^5	$1 \times 10^5 - 3 \times 10^6$	2.1	0.4

[a] Data presented here are for general comparisons only. Since different sampling methods, number of samples, and other procedures were used, the reader should consult the references before using the data for specific planning purposes. Table from Reference 5.

[b] Only ammonia plus nitrate.

[c] Only fecal.

[d] Median values from one sampling station.

[e] Only organic (Kjeldahl) nitrogen.

[f] Only soluble orthophosphate.

Table 2. Comparison of Quality of Combined Sewage for Various Cities[a]

Type of Wastewater, Location, and Year	BOD$_5$ in mg/L		COD in mg/L		DO in mg/L	Suspended Solids in mg/L		Total Coliforms in MPN/100 mL		Total Nitrogen, as N, in mg/L	Total Phosphorus, as P, in mg/L
	Average	Range	Average	Range	Average	Average	Range	Average	Range	Average	Average
Typical Untreated Municipal	200	100–300	500	250–750	—	200	100–350	5×10^7	1×10^7–1×10^9	40	10
Typical Treated Municipal: Primary Effluent	135	70–200	330	165–500	—	80	40–120	2×10^7	2×10^6–5×10^8	35	7.5
Secondary Effluent	25	15–45	55	25–80	—	15	10–30	1×10^3	1×10^2–1×10^4	30	5.0
Selected combined Atlanta, GA, 1969[31]	100	48–540	—	—	8.5	—	—	1×10^7	—	13	1.2[b]
Berkley, CA, 1968–1969[3ac]	60	18–300	200	20–600	—	100	40–150	—	—	—	—

Location											
Brooklyn, NY, 1972[8]	180	86–428	—	—	—	1051	132–8759	—	—	—	1.2b
Bucyrus, OH, 1968–1969[35]	120	11–560	400	13–920	—	470	20–2440	1×10^7	$2 \times 10^5 - 5 \times 10^7$	13	3.5
Cincinnati, OH, 1970[36]	200	80–380	250	190–410	—	1100	500–1800	—	—	—	—
Des Moines, IA, 1968–1969[6]	115	29–158	—	—	—	295	155–1166	—	—	12.7	11.6
Detroit, MI, 1965[2]	153	74–685	115	—	—	274	120–804	—	—	16.3d	4.9
Kenosha, WI, 1970[18]	129	—	464	—	—	458	—	2×10^6	—	10.4d	5.9
Milwaukee, WI, 1969[7]	55	26–182	177	118–765	—	244	113–848	—	$2 \times 10^5 - 3 \times 10^7$	3–24	0.8b
Northhampton, England, 1960–1962[22]	150	80–350	—	—	—	400	200–800	—	—	10d	—
Racine, WI, 1971[18]	119	—	—	—	—	439	—	—	—	—	—
Roanoke, VA, 1969[12]	115	—	—	—	—	78	—	7×10^7	—	—	—

Table 2 (cont'd)

Type of Wastewater, Location, and Year	BOD$_5$ in mg/L		COD in mg/L		DO in mg/L	Suspended Solids in mg/L		Total Coliforms in MPN/100 mL		Total Nitrogen, as N, In mg/L	Total Phosphorus, as P, In mg/L
	Average	Range	Average	Range	Average	Average	Range	Average	Range	Average	Average
Sacramento, CA, 1968–1969[a]	165	70–328	238	59–513	—	125	56–502	5×10^6	$7 \times 10^5 – 9 \times 10^7$	—	—
San Francisco, CA, 1969–1970[b]	49	1.5–202	155	17–626	—	68	4–426	3×10^6	$2 \times 10^4 – 2 \times 10^2$	—	—
Washington, D.C., 1969[c]	71	10–470	382	80–1760	—	622	35–2000	3×10^6	$4 \times 10 – 6 \times 10^6$	3.5	1.0

[a] Data presented here are for general comparisons only. Since different sampling methods, number of samples, and other procedures were used, the reader should consult the references before using the data for specific planning purposes. Table from Reference 5.
[b] Only orthophosphate.
[c] Infiltrated sanitary sewer overflow.
[d] Only ammonia plus organic nitrogen (total Kjeldahl).

project was constructed. The purpose of this project is to improve water quality for recreational activities.[9]

Other cities, i.e., Akron, Ohio; Milwaukee, Wisconsin; Saginaw, Michigan; New York, New York; Chippewa Falls, Wisconsin; and Columbus, Ohio have similar facilities in operation for control of CSO.[5-7]

Costs of storage structures are highly dependent on location. Land, hydrogeological conditions, type structure and construction material, multipurpose usage, and esthetic appearance are primary considerations. Updated capital (excluding real estate), and operation and maintenance costs for storage facilities, are presented in Table 3.[7]

21.3.2 Sedimentation

The time-honored method for removing suspended solids is sedimentation. Storage installations also double as primary sedimentation facilities for flows that exceed storage capacity. Sedimentation has also been used for pretreatment and posttreatment, in addition to it being a dual benefit from storage. Removal efficiencies for CSO storage/sedimentation facilities are usually 30% for BOD_5, and 50% for suspended solids. Significant sedimentation demonstration and prototype project summaries are given in Table 4.[5] Typical pollutant removals are reported in Table 5.[10]

21.3.3 Swirl Overflow Regulator/Concentrator

The swirl regulator/concentrator achieves both quantity and quality control of storm flow simultaneously.[11,12] Anisometric drawing of the device is depicted in Figure 1. The principal mechanism for its dynamic solid-liquid separation ability is secondary fluid motion attained through long-path geometric flow patterns. The low-flow concentrate is diverted to the sanitary sewer system, and relatively clear supernatant overflows the spillway (downshaft), and is diverted to the receiving water with or without subsequent treatment.

Based on a swirl CSO regulator/concentrator prototype demonstration in Syracuse, New York, suspended solids total-mass removals were approximately 50%.[13,14] In addition to the removal obtained by the physical splitting of flows, as with conventional regulators, a 20 to 30% reduction in the suspended solids concentration is attributed to the swirl flow field's separation ability. Swirl concentrators have been hydraulically modeled and are being demonstrated for various other purposes, including grit removal, and stormwater and erosion control.[15]

21.3.4 Fine Screening

Fine screens have been used for various applications including pretreatment, primary treatment, and final polishing. Because they allow relatively high throughput rates, screening units have been used for CSO. Screen openings for

Table 3. Summary of Offline Storage Costs[a]

Location	Storage Capacity, (Mgal)	Drainage Area, (Acres)	Capital Cost,[d] ($)	Storage Cost, ($/Gal)	Cost per Acre, ($/Acre)	Annual O & M Cost, ($/Year)
Akron, OH Milwaukee, WI	1.1	188.5	455,700	0.41	2,420	2,900
Humboldt Avenue Boston, MA	3.9	570	1,774,000	0.45	3,110	51,100
Cottage Farm Detention and Chlorination Station[b] Charles River Marginal Conduit Project New York City, NY	1.3	15,600	6,495,000	5.00	416	80,000
Spring Creek Auxiliary Water Pollution Control Plant	1.2	3,000	9,488,000	7.91	3,160	97,600
Storage	12.39	3,260	11,936,000	0.96	3,660	100,200
Sewer	13.00	—	—	—	—	—
Total	25.39	3,260	11,936,000	0.47	3,660	100,200

Chippewa Falls, WI						
Storage	2.82	90	744,000	0.26	8,270	2,700
Treatment	—	—	189,000	—	2,100	8,000
Total	2.82	90	933,000	0.26	10,370	10,700
Chicago, IL						
Tunnels and pumping	2,998	240,000	870,000,000	0.29	3,630	—
Reservoirs	41,315	—	682,000,000	0.02	2,840	—
Total	44,313	240,000	1,552,000,000	0.04	6,470	—
Total storage treatment	—	—	1,001,000,000	—	4,170	—
Total	44,313	240,000	2,553,000,000	0.04	10,640	8,700,000
Sandusky, OH	0.36	14.86	520,000	1.44	35,000	6,200
Washington, D.C.	0.20	30.0	883,000	4.41	29,430	3,340
Columbus, OH						
Whittier Street	3.75	29,250c	6,144,000	1.64	210	—
Cambridge, MA	0.25	20	320,000	1.28	16,000	14,400

Note: 1 Mgal = 3790 m^3; 1 acre = 0.405 ha; $1/gal = $264/m^3.

[a] ENR 2000. Table from Reference 7.
[b] Estimated values; facilities under design and construction.
[c] Estimated area.
[d] Land costs not included.

Table 4. Summary Data on Sedimentation Basins Combined with Storage Facilities in Operation

Location of Facility	Size, (Mgal)	Type of Storage Facility	Removal Efficiency, as a Percentage		Type of Solids Removal Equipment[a]
			Suspended Solids	BOD$_5$	
Cottage Farm Detention and Chlorination Facility, Cambridge, MA	1.3	Covered concrete tanks	45	Erratic	Manual washdown
Chippewa Falls, WI	2.8	Asphalt paved storage basin	18–70	22–74	Solids removal by street cleaners
Whittier Street	4.0	Open concrete tanks	15–45	15–35	Mechanical wash-down
Alum Creek	0.9	Covered concrete tanks	NA[b]	NA	Mechanical wash-down
Humboldt Ave., Milwaukee, WI	4.0	Covered concrete tanks	NA	NA	Resuspension of solids by mixers

| Spring Creek, Jamaica Bay, New York, NY | 10.0 | Covered concrete tanks | NA | NA | Traveling bridge hydraulic mixers |

Note: 1 Mgal = 3790 m^3.

[a] All facilities store solids during storm event and clean sedimentation back in when flows to the interceptor can handle the wash water and solids.

[b] NA = not available.

Table 5. Pollutant Removal for Various
Constituents by Sedimentation[10]

Pollutant	Average Removal, as a Percentage
Metals and heavy metals[a]	
Copper	24.1
Chromium	32.3
Nickel	26.6
Zinc	27.2
Lead	30.6
Iron	16.6
Cadmium	38.8
Calcium	19.2
Magnesium	23.5
Sodium	18.5
Potassium	23.5
Mercury	8.4
Nitrogen[b]	
Ammonia	22.1
Organic	50.5
Total kjeldahl	38.4
Nitrate	15.4
Nitrite	0
Phosphorus[b]	
Total	22.2
Ortho	6.7
Other constituents[b]	
COD	34.4
TOC	21.3
Oil and grease[c]	11.9

[a] Average of ten samples.
[b] Average of two to three samples.
[c] Average of six samples.

microstrainers range from 23 to 100 μ, and for the fine mesh type from 100 to 1700 μ. Removal efficiency is primarily a function of screen aperture dimension and flow suspended solids characteristics, and increases as the suspended solids grain size and concentrations increase and aperture dimension decreases. Solids removal results from two basic treatment mechanisms: (1) direct straining by the screen, (2) filtering of smaller particles by the mat, deposited on the screen surface by straining. Microstrainers and fine-mesh screens remove from 10% to

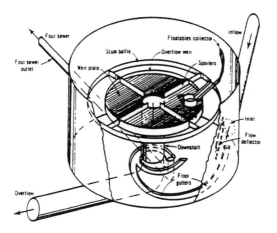

Figure 1.   Isometric view of swirl regulator/concentrator.

99% of the suspended solids and from 10% to 70% of the BOD$_5$. A summary of operation and cost (1980) data is presented in Table 6.[5]

21.3.5 Filtration

In the area of physical treatment, filtration is one step finer than screening. Demonstrations made on filtration for treating urban stormwater and CSO have shown the most successful type to be the high-rate dual-media filter using anthracite coal over sand. A preliminary (420μ) screening process was used upstream of the filters to extend the treatment run time before backwashing. It was found that suspended solids removal increased as influent suspended solids concentration increased, and decreased as hydraulic loading increased.

Removal efficiency for the filter unit was about 65% for suspended solids, 40% for BOD$_5$, and 60% for chemical oxygen demand (COD). The addition of a polyelectrolyte increased the suspended solids removal to 94%, the BOD$_5$ removal to 65%, and the COD removal to 65%. The length of filtration run averaged 6 h at loading rates of 24 gpm/ft^2 (1400 m/d).[16]

21.3.6 Biological Treatment Processes

Biological treatment of wastewater produces an effluent of high quality. These processes are generally categorized as secondary treatment processes capable of removing between 80% and 95% of the BOD$_5$ and suspended solids from sanitary sewage.[7] When biological treatment processes are used for stormwater treatment, removal efficiencies are lower and controlled to a large degree by hydraulic and organic loading rates. As previously mentioned, most

Table 6. Characteristics of Various Types of Screens

Characteristic	Microstrainer	Drum Screen	Rotary Fine Screen	Hydraulic Sieve
Principal use	Main treatment	Pretreatment to other devices and main treatment	Pretreatment to other devices and main treatment	Pretreatment to other devices
Approximate removal efficiency, as a percentage				
BOD	50	15	15	—
Suspended solid	70	40	35	—
Land requirements, (ft²/mgd)	~15–20	5–20	24–62	20
Cost, ($/Mgal)[a]	12,000	4,800	8,000	5,600
Can be used as a dry-weather flow polishing device?	Yes	No	No	No

	Possible, with controls	Possible, with controls	Possible, with controls	No controls needed
Automatic operation?	Possible, with controls	Possible, with controls	Possible, with controls	No controls needed
Able to treat highly varying flows?	Yes	Yes	Some limitation	Yes
Removes only particulate matter?	Yes	Yes	Yes	Yes
Requires special shutdown and startup regimes?	Yes	Some	Some	No
Screen life with continuous use?	7–10 yr	10 yr	1,000 h	20 yr
Uses special solvents in back-wash water?	No	No	Yes	No
High solids concentrate volume, as a percentage of total flow	0.5–1.0	0.5–1.0	10–20	<0.5

Note: 1 sq ft/mgd = 4.65 ha/m^3/d; $1 mgd = $0.000264/m^3/d.

[a] Based on a 25-mgd (95,000 m^3/d) plant capacity, land costs not included.

Table 7. Typical Wet-Weather BOD and Suspended Solids Removals for Biological Treatment Processes[7]

Biological Treatment Process	Expected Range of Pollutant Removal, as a Percentage	
	BOD	Suspended Solids
Contract stabilization	70–90	75–95
Trickling filters	65–85	65–85
Rotating biological contractors[a]	40–80	40–80

[a] Removal reflects flow ranges from 30 to 10 times dry weather flow.

biological systems are extremely susceptible to overloading conditions and shock loads when compared to physical treatment processes. However, rotating biological contractors (RBC) have achieved high removals at flows 8 to 10 times dry weather design flows.[17] Typical pollutant removals for contact stabilization, trickling filters, and RBC are presented in Table 7 for wet-weather loading conditions.[7] These processes include primary and final clarification.

Because of the limited ability of biological systems to handle fluctuating and high hydraulic and organic solids shock loads, storage facilities preceding the biological processes must be considered.

21.3.7 Activated Carbon Adsorption

The role of the carbon adsorption step is to remove dilute soluble refractory, or residual organics, or both, from the wastewater. Carbon-contacting systems generally employ granular activated carbon. The wastewater is passed either downward or upward through the carbon bed. An upflow expanded bed type is commonly used for wastewater treatment. The influent is introduced into the bottom of the column and flows up through the column at rates of 2 to 8 gpm/ft^2 (120 to 470 m/d). Exhausted carbon is removed from the bottom of the column and an equivalent quantity of fresh, or regenerated carbon, or both, is added to the top of the column. Periodical backwashing may be needed to remove excess biological growth and solids.

Organic demand removal capacity of carbon is about 0.5 lb COD/lb granular carbon (0.5 kg COD/kg granular carbon).[5] This is approximately equal to a requirement of 500 lb (225 kg) of activated carbon per million gallons of wastewater treated. When activated carbon is used for polishing or tertiary treatment, removal efficiencies for BOD_5 and suspended solids are above 95%.

A new development that employs powdered in lieu of granular activated carbon, which eliminates the need for the contact column, was demonstrated for

the EPA for treating CSO, in Albany, New York.[18] This system consisted of a
0.10-mgd (380 m³/d) trailer-mounted pilot plan where both powdered carbon and coagulants were added in a static mixing/reaction pipeline, and the
resultant coagulated matter was flocculated downstream and separated by tube
settlers.

21.3.8 Disinfection

Disinfection of urban stormwater is generally practiced at treatment facilities
to control pathogenic microorganisms. At most stormwater and CSO control
installations, disinfection has been accomplished by applying conventional
technology supplemented by high-rate processes or on-site disinfectant generation. Studies have been made on high dosages, high-energy static and dynamic
mixing, more rapid oxidants (e.g., ozone, ultraviolet radiation, and chlorine
dioxide), or two-stage disinfection (using chlorine and chlorine dioxide), or
both, to enhance the rate of bacterial kill. Successful attempts towards high-rate
disinfection have been reported at the Philadelphia, Pennsylvania, Onondaga
County, New York, and Monroe County, New York, demonstration sites.[19-21]
The effectiveness of a disinfection system is influenced by contact time, type
of disinfectant, dosage, mixing intensity, pH, protective suspended matter in the
flow, and temperature. The contact time required for the process is inversely
proportional to the concentration of the disinfectant and water temperature. For
instance, it required five times as much contact time to obtain a 99% kill at 5°C,
as it does to obtain the same kill at 30°C, at the same disinfectant dosage.
Temperature is an important consideration here because storm runoff is more
prone to temperature fluctuation. Therefore, stormwater disinfection must be
flexible and also capable of automatic operation to handle intermittent and
varying storm flows.

21.4 URBAN STORMWATER RECLAMATION

The basic treatment and control methods indicated in the previous section
have been examined essentially as singular processes; however, system
optimization should be considered to enhance overall treatment effectiveness
and stormwater reuse potential. This section deals with integrated treatment
systems to produce four levels of effluent for differing water quality classifications.

A typical flow diagram for several advanced physical-chemical treatment
systems is shown in Figure 2. Each treatment system produces a different degree
of effluent water quality, except for sedimentation and fine screening high-rate
filtration which result in similar treatment efficiencies. The flow diagram also
indicates the water quality classifications associated with different water uses.
Class AA is intended for high quality application, such as steam generation boiler
feed. Class A is intended for routine industrial process supply, which has lower

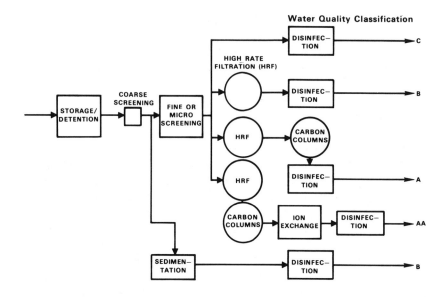

Figure 2.   Typical process flow diagram for several advanced physical/ chemical treatment systems.

dissolved mineral removal requirements than Class AA. Class B can be used for industrial cooling and recreational water for fishing. Additional nutrient removal may be warranted. Finally, Class C is intended for lawn irrigation, fire protection, and esthetic ponds.

Table 8 indicates maximum concentrations of selected quality parameters.[22] The water demands of each use must be evaluated with respect to the four aforementioned water quality levels defined in Table 8. The treatment systems described in the foregoing paragraphs are only intended to be representative of general conditions. Specific unit process selection and system design should only be determined after suitable investigation of the quality and quantity characteristics of the stormwater at the site and laboratory/pilot-scale evaluations.

21.1 HYPOTHETICAL CASE STUDY

A case study for a hypothetical industrial complex is presented to illustrate an engineering economic analysis of urban stormwater reclamation. Three water quality classification levels (A, B, and C, as shown in Table 9) were used to represent a range of industrial applications.Table 10 presents typical water

Table 8. Maximum Concentration of Selected Pollutant by Reuse Category[22]

Constituent, (mg/L), Unless Indicated	AA	A	B	C
	Required Water Quality, Maximum Concentration[a]			
	Water Quality Classification			
Ammonia (NH$_3$)	0.5	0.5	0.5	0.5
Arsenic	0.01	0.05	0.05	0.05
alcium	0.5	75.0	75.0	75.0
Chloride	50.0	250.0	250.0	250.0
Chromium (hexavalent)	0.05	0.05	0.05	0.05
Copper	1.0	1.0	1.5	1.5
Cyanide	0.01	0.2	0.2	0.2
Fluoride	1.5	3.0	3.0	3.0
Iron	0.1	0.3	0.3	0.3
Lead	0.05	0.1	0.1	0.1
Magnesium	0.05	150.0	150.0	150.0
Manganese	0.05	0.1	0.5	0.5
Nitrate (as NO$_3$)	45.0	50.0	50.0	50.0
Oxygen, dissolved (minimum)	5.0	5.0	4.0	4.0
Sulfate	50.0	200.0	400.0	400.0
Total solids	150.0	500.0	500.0	1500.0
Zinc	5.0	15.0	15.0	15.0
Coliform (MPN/100 mL)	1	70	240	240
pH (units)	7.0	6.0	6.0	6.0
Color (units)	15	20	30	30
Turbidity (units)	0–3	3–8	8–15	15–20
Suspended solids	—	—	10.0[b]	30.0[b]
Phosphates	1.0	1.0	1.0	1.0

[a] Based upon maximum concentrations allowed by the United States Public Health Service, the World Health Organization, and the Water Quality Standards of the State of Maryland.

[b] Higher suspended solids are permitted by various water quality standards. Limit based on sediment control and water contact recreation.

Table 9. Water Demands and Quality Classifications
(Hypothetical Case Study)

| Use Category | Water Demand | | Water Quality Classification |
	In Millions of Gal/D (m³/d)	In Millions of Gal/Year (Millions of m³/year)	
Lawn irrigation	0.39	9.4	C
	(1,480)	(0.356)	
Cooling makeup	5.0	1,500	B
	(19,000)	(5.69)	
otable	0.89	267	City water
	(3,370)	(1.01)	
Process	1.61	483	A
	(6,100)	(1.83)	
Overflow disinfection	9.11	20	C
	(34,500)	(0.076)	

quality requirements for cooling tower makeup and process supply. These are assumed to be hypothetical case study requirements. Three water supply sources combinations, including reclaimed urban stormwater, or city water, or both, were considered as potential alternatives for this case study.

21.5.1 Stormwater Reclamation System

The stormwater reclamation system includes a storage reservoir and treatment facilities. The basic purpose of the system is industrial water supply, but the system also provides the multibenefits of enhanced drainage control, water pollution abatement, and improved local esthetics. In other words, the storage basin functions as a water supply reservoir, flow attenuator, pollutant load attenuator or separator, and esthetic pond. For pollution control, the reservoir should be designed and operated to prevent degradation of downstream receiving-water quality. For water supply, the significant design criteria are industrial water demand rate, storage capacity, and the reliability with which the demand can be satisfied.

In studying storage requirements for the case study, the basin drainage locations (or catchment areas) were first analyzed to determine the industrial plant yield vs volume/capacity relationships, assuming 100% reliability over a representative 5-year buildup period of stormwater inflow volume. A mass-flow accumulation diagram (volumetric flow accumulation over a long time) with the known industrial process demand rate was used for this purpose.

Table 10. Water Quality Requirements
(Hypothetical Case Study)

Parameter, in mg/L	Cooling Makeup	Process Supply
Total dissolved solids	1500	500
Suspended solids	25–100	<10
pH[a]	6.9–7.5	7
Turbidity[b]	20	5
Calcium	50	50
Iron	0.5	0.3
BOD_5	25	2.0
COD	75	5
Sulfate	200	5

[a] In pH units.
[b] In Jackson turbidity units.

Next, effluent water quality was analyzed in terms of system design and operating requirements. Finally, treatment process design criteria and capacities were developed which met all three plant water use (water classes A, B, and C) requirements. Detailed methodology and procedures for hydrological and process analyses are examined in other publications.[5,7,22,23]

Fine-mesh screening has had relatively wide use for storm flow treatment having been adopted for both the main process and for pretreatment in such processes as high-rate filtration and dissolved air flotation.[5,7,15] High-rate filters hold promise for stormwater treatment based on large-scale pilot demonstrations.[5,7,16,24] Carbon adsorption is usually employed as a tertiary polishing step if a higher water quality is desired, and also has a demonstrated ability to treat stormwater.[5,18]

The proposed treatment system contains these processes and is depicted in Figure 3. The design flows and criteria of the unit processes are contained in Table 11.

The capital costs were adjusted using the national average EPA WQO-Sewage Treatment Plant Index of 303.1 (June 1978). Annual costs were based on a 25-year amortization period at 10% interest and 300 days/year of plant operation.

The proposed storage reservoir is of lined earthen construction and is 18-ft deep with a 2.5:1 interior slope, a 3:1 exterior slope, and a 16-ft top width of levee. The reservoir is esthetically designed to harmonize with the local environment. Table 12 summarizes the capital, O & M, and total annual costs for the proposed stormwater reclamation treatment facilities using unit costs from a cost-estimating manual.[25]

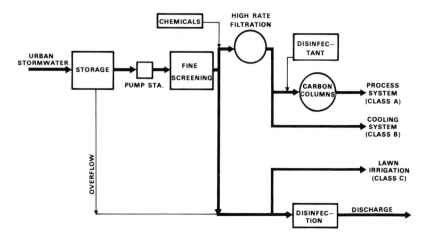

Figure 3.   Proposed treatment system flow diagram.

Table 11. Design Criteria for Storage-Treatment Facilities
(Hypothetical Case Study)

Unit	Design Flow Rate		Design Criteria
	In Millions of Gal/D	**In Gal/Min.**	**Design Criteria**
Flow control/ treatment storage basin	200[a]		30 days
Hgh-rate disin- fection units	10	7000	15 min.
Class C, B, & A pumping station	7.2	5000	—
fine-mesh screens	7.2	5000	50 gpm/sq ft
Cass B & A high-rate filters	6.6	4630	16 gpm/sq ft
Class A carbon columns	1.6	1130	5 gpm/sq ft 30 min.

[a]   In millions of gallons; 1 mgd = 3,790 m³/d; 1 gpm/sq ft = 58.7 m/d.

Table 12. Capital and Annual Costs for Urban
Stormwater Treatment (Hypothetical Case Study)[a]

Process/ Equipment	Capital Cost, ($)	Annual Cost, ($)		
		Amortization[b]	O & M	Total
Flow control/ treatment				
Storage basin	3,600,000	396,000	10,000	406,000
High-rate disin- fection units	166,000	18,000	117,000	135,000
Class C, B, & A				
Pump station	975,000	107,000	90,000	197,000
Fine screening units	208,000	23,000	25,000	48,000
Class B & A				
High-rate filtration units	1,200,000	132,000	36,000	168,000
Chemical feed equipment and building	224,000	25,000	100,000	125,000
Class A				
Carbon columns	1,300,000	143,000	90,000	233,000
Disinfectant feed system	120,000	13,000	100,000	113,000

[a] Unit cost from Reference 25.
[b] $n = 25$ years; $i = 10\%$.

Table 13 presents the summary of total annual costs of water supply for each alternative.

Based on the aforementioned analysis, Alternative 3, utilizing a combination of storm and city water supply, offers a clear savings of approximately $900,000/year over Alternative 1, which utilizes city water only. Even Alternative 2, which is mainly dependent on stormwater supply is $800,000/year less than the exclusive city water supply alternative. Further savings are realized when the multipurposes of pollution control, drainage, and esthetics are considered.

Table 13. Estimated Total Annual Costs of Water Supply (Hypothetical Case Study)

Purpose	Source	Annual Flow, (Mgal)	Unit cost, ($/1000 gal)	Annual Costs, ($)
		Alternative 1		
Irrigation	City water	94.0	0.80	75,000
Cooling	City water	1,500.0	1.05	1,575,000
Process	City water	483.0	1.20	9,580,000
Potable	City water	267.0	0.80	214,000
TOTAL				2,440,000
		Alternative 2		
Irrigation	Class C (stormwater)	94.0	0.34	32,000
Cooling	Class B (stormwater)	1500.0	0.48	720,000
Process	Class A (stormwater)	483.0	1.44	696,000
Potable	City water	267.0	0.80[a]	214,000
TOTAL				1,662,000
		Alternative 3		
Irrigation	Class C (stormwater)	94.0	0.34	32,000
Cooling	Class B (stormwater)	1,500.0	0.48	720,000
Process	City water	483.0	1.20[b]	580,000
Potable	City water	267.0	0.80[a]	214,000
TOTAL				1,546,000

Note: 1 Mgal = 3790m^3; $1/1000 gal = $0.264/m^3.

[a]   Average water consumption charge rate in Middlesex County, NJ (June 1978).

[b]   Additional treatment required for control of biological growth, corrosion, or scaling or both.

21.6 CONCLUSIONS

This hypothetical case study provides evidence that the reclamation of urban stormwater for industrial subpotable water supply is technically feasible and economically attractive when compared to the city water source. In addition, other important benefits, such as reduction of pollutant discharges, drainage control, creation of recreational and esthetic ponds, groundwater recharge, and improvement and preservation of ecology in urban areas will be achieved.

REFERENCES

1. Schmidt, C.J. "Current municipal wastewater reuse practices-research needs for the potable reuse of municipal wastewater," U.S. EPA Report-600/9-75-007 (1975).
2. Dean, R.B., and J.J. Convery. "Wastewater treatment technology for potable reuse-research needs for the potable reuse of municipal wastewater," U.S. EPA Report-600/9-75-007 (1975).
3. Garrison, W.E., and R.P. Miele. "Current trends in water reclamation technology," *J. Am. Water Works Assoc.* 69:7 (1977).
4. American Water Works Association. "Water reuse highlights" (1978).
5. Lager, J.A., and W.G. Smith. "Urban stormwater management and technology: an assessment," U.S. EPA Report-670/2-74-040, NTIS No. PB 240 687 (1974).
6. Field, R., et al. "Urban runoff pollution control program overview: FY76," U.S. EPA Report-600/2-76-095, NTIS No. PB 252 223 (1976).
7. Lager, J. A., et al. "Urban stormwater management and technology: update and users' guide," U.S. EPA Report No. EPA 600/8-77-014, NTIS No. PB 275 654 (1977).
8. "Cottage farm combined sewer detention and chlorination station, Cambridge, Massachusetts," U.S. EPA Report-600/2-77-046, NTIS No. PB 263 292 (1976).
9. Metropolitan District Commission, Commonwealth of Massachusetts. "Environmental assessment statement for Charles River marginal conduit project in the cities of Boston and Cambridge, Massachusetts," (1974).
10. New York City Environmental Protection Administration. "Spring creek auxiliary water pollution control plan operational data," January 1974 to January 1976 (1976).
11. Sullivan, R.H., et al. "The swirl concentrator as a combined sewer overflow regulator facility," U.S. EPA Report-R2-72-008, NTIS No. PB 214 687 (1972).

12. Sullivan, R.H., et al. "Relationship between diameter and height for design of swirl concentrator as a combined sewer overflow regulator," U.S. EPA Report-670/2-74-039, NTIS No. PB 234 646 (1974).

13. Drehwing, F.J., et al. "Disinfection/treatment of combined sewer overflows Syracuse, New York," U.S. EPA Report-600/2-79-134, NTIS No. PB 80-113459 (1979).

14. Field, R.I. "Treatability determinations for a prototype swirl combined sewer overflow regulator/solids-separator, proceedings of the urban stormwater management seminars, 1975," U.S. EPA Report No. WPD 03-76-04, NTIS No. PB 260 889 (1974).

15. Field, R.I., and H. Masters. "Swirl device for regulating and treating combined sewer overflows," U.S. EPA Report-625/2-77-012 (1977).

16. Nebolsine, R., et al. "High rate filtration of combined sewer overflows," U.S. EPA Report-11023EYI04/72, NTIS No. PB 211 144 (1972).

17. Welsh, F.L., and D.J. Stucky. "Combined sewer overflow treatment by the rotating biological contractor process," U.S. EPA Report-670/2-74-050 (1974).

18. "Physical-chemical treatment of combined and municipal sewage," Battell Northwest Laboratories, U.S. EPA Report-R2-73-149, NTIS No. PB 219 668 (1973).

19. Glover, G.E., and G.R. Herbert. "Microscreening and disinfection of combined sewer overflow, phase II," U.S. EPA Report-R2-73-124, NTIS No. PB 219 879 (1973).

20. Moffa, P.E., et al. "Bench-scale high-rate disinfection of combined sewer overflows with chlorine and chlorine dioxide," U.S. EPA Report-670/2-75-021, NTIS No. PB 242 296 (1975).

21. Geisser, D.F., et al. "Design optimization of high-rate disinfection using chlorine and chlorine dioxide," *J. Water Pollut. Control Fed.* 51:2 (1979).

22. Mallory, C. W. "The beneficial use of stormwater," U.S. EPA Report-R2-73-139 (1973).

23. Lindsey, R.K., et al. *Applied Hydrology* (New York: McGraw-Hill Book Co., Inc., 1949).

24. Innerfeld, H., et al. "Dual process high rate filtration of raw sanitary sewage and combined sewer overflows," U.S. EPA Report-600/2-79-015, NTIS No. PB 296 626 (1979).

25. Benjes, H.H., Jr. "Cost estimating manual-combined sewer overflow storage and treatment," U.S. EPA Report-600/2-76-286, NTIS No. PB 266 359 (1976).

STORMWATER MODELS AND MONITORING

22 PREDICTION AND ESTIMATION OF STORMWATER RUNOFF CONTROL BY THE COMBINED STORMWATER INFILTRATION FACILITIES WITH SIMULATION MODEL

22.1 INTRODUCTION

Many stormwater infiltration facilities, for example porous asphalt pavement, infiltration stormwater inlets, and infiltration stormwater pipes, have been constructed to appropriately control urban storm runoff in Japan. In addition, many sites contain more than one facility in order to achieve effective stormwater control in urban areas.

It is difficult to predict the volume of stormwater runoff that is reduced by the use of stormwater infiltration facilities. This is in contrast to the stormwater reservoirs for flood control whose reservoir volumes can be predicted easily. For this reason, the stormwater infiltration volume is predicted based on the infiltration capacity obtained by infiltration tests on the same type of stormwater infiltration facility, but constructed at other sites.

In this paper, we have measured the infiltration capacities of several types of combined stormwater infiltration facilities constructed at an experimental field. We have also researched effective combinations of stormwater infiltration facilities for stormwater runoff control; investigated the mechanism of stormwater infiltration; and built a simulation model to predict the volume of stormwater infiltration at the combined stormwater infiltration facilities.

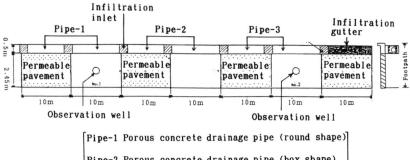

Figure 1.   Layout of our experimental infiltration facilities.

22.2 EXPERIMENTAL FACILITIES AND METHOD

22.2.1 Outline of Experimental Infiltration Facilities

The reduction levels of stormwater runoff volume, by combining stormwater infiltration facilities constructed for experimentation at the site of the treatment plant, were measured. The following stormwater infiltration facilities were constructed at the experimental field location:

1. permeable pavement for footpath: 4 pavements
2. infiltration trench: 3 types (2 lengths each)
3. infiltration roadside gutter (L type): 1 length
4. infiltration street inlet: 7 inlets

In addition, there are two infiltration wells to measure ground infiltration capacity.

Figure 1 shows the layout of our experimental infiltration facility. The experimental facility is a footpath (L = 70m, W = 2.95m) and each experimental infiltration trench is constructed under the roadside gutter (W = 0.5m). Porous asphalt pavements and usual asphalt pavements are constructed alternately. The area of each porous asphalt pavement is 24.5 m^2 (L = 10.0m, W = 2.45m) and the length of each infiltration trench is 8.7 m.

The combinations of the infiltration facilities are as follows:

- Type 1: permeable pavement + porous concrete drainage pipe (round shape)
- Type 2: permeable pavement + porous concrete drainage pipe (box shape)
- Type 3: permeable pavement + vinyl chloride pipe with holes (egg shape)

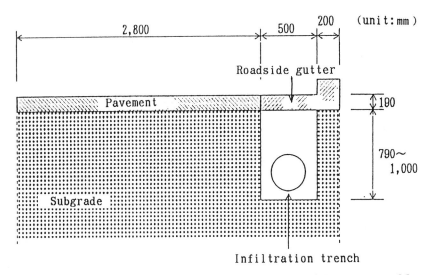

Figure 2. Cross-section of infiltration trench with unpermeable pavement.

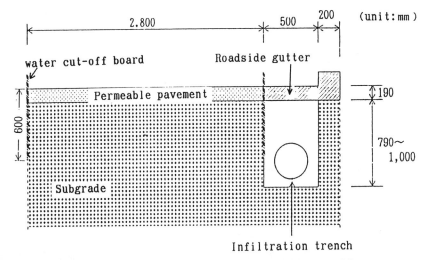

Figure 3. Cross-section of infiltration trench with permeable pavement.

Figures 2 and 3 show the cross sections of the combined infiltration facilities. The water cutoff boards are constructed between the base courses under the permeable pavements and the infiltration trenches. The vertical length of the water cutoff boards is 0.59 m from the ground surface.

A cross-section of the permeable pavement is shown in Figure 4. Cross sections of infiltration facilities are shown in Figures 5 to 7.

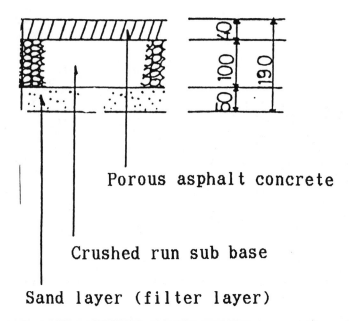

Porous asphalt concrete

Crushed run sub base

Sand layer (filter layer)

Figure 4. Cross-section of the permeable pavement.

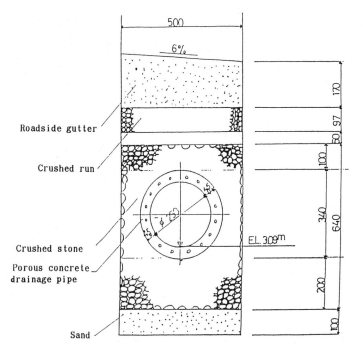

Figure 5. Cross-section of porous concrete drainage pipe (round shape).

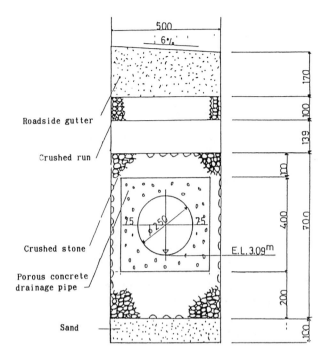

Figure 6.   Cross-section of porous concrete drainage pipe (box shape).

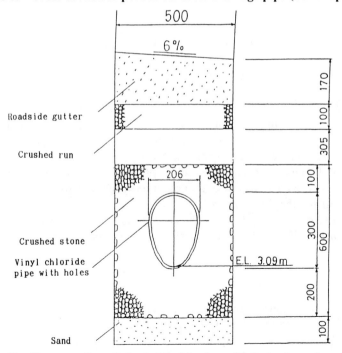

Figure 7.   Cross-section of vinyl chloride pipe with holes (egg shape).

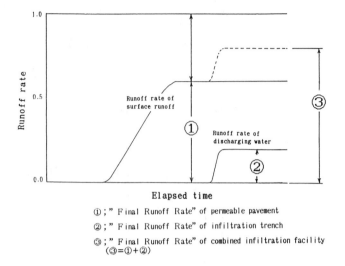

Figure 8.   Changes of surface runoff and discharging water at the combined infiltration facilities.

22.2.2 Process of Experiments

Ground Infiltration Capacity Test

An acrylic resin pipe (Ø 0.12 m) was installed to measure the ground infiltration capacity of the experimental site by a constant head test.

Combined Infiltration Facilities

Water was sprinkled on the surface of the permeable pavements to simulate rainfall, and the time required for surface runoff to begin and the volume of surface runoff water were measured. Figure 8 shows the changes in surface runoff and discharging water with time when the sprinkled rainfall intensity is constant. When the amount of surface runoff water from the permeable pavement and discharged water from the infiltration trench is nearly constant, we call the runoff rate the *Final Runoff Rate* (F_r).

Initial Infiltration Height, which is volume of stormwater infiltration until runoff begins, is calculated with the following equation:

$$F_I = (R \cdot T_i) / A \tag{1}$$

in which, F_I is the initial infiltration height (mm), R is volume of sprinkled or poured water per minute (l/min), T_i is the time taken for stormwater runoff to be started (min), and A is the area of permeable section (m^2).

Final Infiltration Capacity is calculated with the following equation when the amount of surface runoff water from the permeable pavement and discharged water from the infiltration trench is nearly constant:

$$F_N = (R - Q_S) / A \cdot 60 \qquad (2)$$

in which, F_N is the final infiltration capacity (mm/h), Q_S is the amount of surface runoff water from the permeable pavement that is discharged   from the infiltration trench per minute (l/min).

22.3 EXPERIMENTAL RESULTS

22.3.1 Infiltration Capacity of Experimental Site

Thirty-five constant head tests were conducted. The infiltration capacity of the experimental site was about 900 mm/h and ranged from 300 mm/h to 1500 mm/h.

22.3.2 Initial Infiltration Height and Final Infiltration Capacity

Figure 9 shows an example of the infiltration test at the combined infiltration facilities. Ranges and average values of initial infiltration height and final infiltration capacity were as follows:

		Range	Average
Type 1:	F_I	17–54 mm	32 mm
	F_N	8–16 mm/h	12 mm/h
	F_r	0.53–0.88 mm	0.71
Type 2:	F_I	11–56 mm	32 mm
	F_N	12–14 mm/h	13 mm/h
	F_r	0.78–0.86 mm	0.82
Type 3:	F_I	24–43 mm	33 mm
	F_N	25–33 mm/h	29 mm/h
	F_r	0.51–0.59 mm	0.55

Based on our experiments, we placed the combined infiltration facilities in order by the final infiltration capacity. The order was as follows:

- Type 3: permeable pavement + vinyl chloride pipe with holes (egg shape)
- Type 2: permeable pavement + porous concrete drainage pipe (box shape)

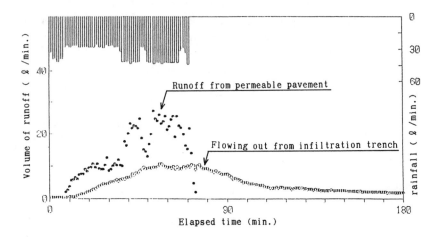

Figure 9. **Result of infiltration test at the combined infiltration facilities.**

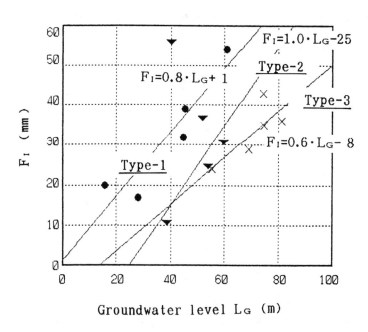

Groundwater level L_G (m)

Figure 10. **Relation between the groundwater level and initial infiltration height.**

- Type1: permeable pavement + porous concrete drainage pipe (round shape)

Because the capacity is very much influenced by the groundwater level and the water content of the soil at the site, further investigations are necessary before any conclusions can be drawn on infiltration capacity.

22.3.3 Influence Factors for Infiltration Capacity of Infiltration Facility

The following five factors are the main factors influencing the infiltration capacity of an infiltration facility:

1. temperature
2. permeability coefficient of the ground
3. groundwater level
4. nature of the soil
5. structural characteristics of infiltration facility

Groundwater Level

In our experimental field, groundwater level is constantly high and its depth from the ground surface varies from 1.5 to 0.5 m. Figure 10 shows the relationship between the groundwater level and the initial infiltration heights of each combination of permeable pavement and infiltration trench. In this paper, we have represented groundwater levels by the distance from the bottom of the gravel layer laid under the porous asphalt pavement. The initial infiltration height in each combination is varied by the groundwater level. Figure 10 shows that they have a simple proportional relationship.

The regression equation between the distances from the bottom of the gravel layer to the groundwater surface and the initial infiltration height were calculated. The following regression equations were obtained at all combinations of infiltration facilities:

$$\text{Type 1: } F_1 = 0.8\, L_G + 1 \quad (r = 0.93) \tag{3}$$

$$\text{Type 2: } F_1 = 1.0\, L_G - 25 \quad (r = 0.79) \tag{4}$$

$$\text{Type 3: } F_1 = 0.6\, L_G - 8 \quad (r = 0.78) \tag{5}$$

where L_G is the distance from the bottom of the gravel layer to the groundwater surface (m).

Initial infiltration height (F_1) of Type 1 is highest in our experimental infiltration trenches when groundwater level ranges from –0.5 to –1.5m.

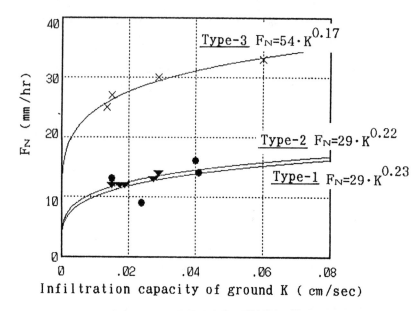

Figure 11. Relation between the infiltration capacity of ground and final infiltration capacity.

Ground Infiltration Capacity

Figure 11 shows the relation between the infiltration capacity of the ground and the final infiltration capacities of the combined infiltration facilities. The final infiltration capacity in each combination is varied by the infiltration capacity of the ground. Each final infiltration capacity is related to the infiltration capacity of ground by the following equations:

$$\text{Type 1: } F_N = 29 \ K^{0.23} \qquad (r = 0.45) \qquad\qquad (6)$$

$$\text{Type 2: } F_N = 29 \ K^{0.22} \qquad (r = 0.91) \qquad\qquad (7)$$

$$\text{Type 3: } F_N = 54 \ K^{0.17} \qquad (r = 0.97) \qquad\qquad (8)$$

in which K is the infiltration capacity of the ground at the experiment of combined infiltration facilities (cm/sec). With these equations we can predict the final infiltration capacity of the combined infiltration facilities. If we measure the final infiltration capacities of combined infiltration facilities under the same ground condition, then Type 3 exhibits the highest infiltration capacity. The infiltration capacity of Type 3 is about 15 mm/h higher than that of Type 2 in any ground infiltration capacities.

22.4 MECHANISM OF INFILTRATION AND SIMULATION MODEL

22.4.1 Mechanism of Infiltration

The mechanism of infiltration at the combined permeable pavement and infiltration trench are considered as follows:

Permeable Pavement

1. Because the infiltration capacity of the permeable pavement is higher than that of the subgrade of permeable pavement, stormwater falling on the surface of permeable pavement is reserved in the permeable pavement.
2. If rainfall continues with heavy rainfall intensity exceeding the Final Infiltration Capacity of the permeable pavement and the permeable pavement is saturated with the infiltrated stormwater, stormwater runoff occurs on the surface.
3. Infiltration velocity changes with the wetting condition at the subgrade of permeable pavement.
4. Part of the infiltrated stormwater for the subgrade of permeable pavement infiltrates horizontally into the infiltration trench installed under one side of the permeable pavement. The volume of the horizontal infiltration from the subgrade of permeable pavement is changed by the volume of the temporarily stored stormwater at the subgrade of permeable pavement.

Infiltration Trench

1. Part of the horizontally-infiltrated stormwater from the subgrade of permeable pavement reinfiltrates through the infiltration trench into the soil. The remaining infiltrated stormwater is stored temporarily at the gravel section of the infiltration trench.
2. If the water level of the temporarily-stored stormwater reaches the inner bottom of the permeable pipe at the gravel section of the infiltration trench, then the temporarily stored stormwater flows out.

22.4.2 Simulation Model

Our simulation model consists of the following three parts.

1. permeable pavement model
2. subgrade layer under the permeable pavement model
3. infiltration trench model

Our model represents the infiltration condition from the permeable pavement to the subgrade layer under the permeable pavement and from the subgrade layer under the permeable pavement to the infiltration trench.

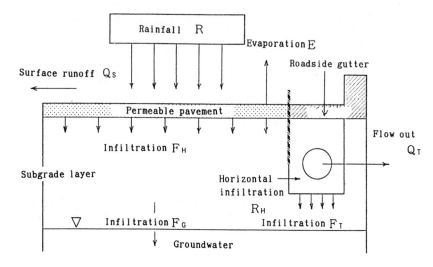

Figure 12. Concept in our simulation model.

The concept in our simulation model is shown in Figure 12.

Permeable pavement model

Effective Rainfall

$$R_E(t) = (R(t) - E(t)) \cdot A_H \qquad (9)$$

Infiltration

$$F_H(t) = S_H(t) \cdot A_H \qquad (10)$$

Surface Runoff

$$V_H{}' = V_H(t-1) + R_E(t) - F_H(t) \qquad (11)$$

$$Q_S(t) = 0 \qquad (V_H{}' < M_H; \text{ permeable pavement is not saturated}) \quad (12)$$

$$Q_S(t) = V_H{}' - M_H \qquad (V_H \geq M_H; \text{ permeable pavement is saturated}) \quad (13)$$

Temporarily Stored

$$V_H(t) = V_H{}' - Q_S(t) \qquad (14)$$

where:

$R_E(t)$ =   *volume of effective rainfall ($l/\Delta t$)*

$R(t)$ =   *rainfall ($mm/\Delta t$)*

$E(t)$ =   *evaporation ($mm/\Delta t$)*

A_H =   *area of permeable pavement (m^2)*

$F_H(t)$ =   *volume of infiltration water from permeable pavement to the subgrade layer ($l/\Delta t$)*

$S_H(t)$ =   *infiltration velocity at the permeable pavement ($mm/\Delta t$)*

$V_H(t)$ =   *volume temporarily stored in the permeable pavement (l)*

M_H =   *maximum volume of temporarily stored in the permeable pavement (l)*

$$M_H = M_{H\upsilon} \cdot A_H \tag{15}$$

where:

M_{HU} =   *unit volume of void of the permeable pavment (mm)*

$Q_S(t)$ =   *volume of surface rainfall runoff water*

Subgrade layer under the permeable pavement model

Horizontal Infiltration

$$V_G{}' \quad = V_G(t-1) + F_H(t) \tag{16}$$

$$R_H(t) = 0 \qquad \left(V_G{}' < M_G\right) \tag{17}$$

$$R_H(t) = \left(V_G{}' - M_G\right) \cdot f \quad \left(V_g{}' \geq M_G\right) \tag{18}$$

Infiltration

$$F_G(t) = S_G(t) \cdot A_H \tag{19}$$

Temporarily Stored

$$V_G(t) = V_G' - R_H(t) - F_G(t) \tag{20}$$

where,

$V_G(t)$ = *volume of temporary stored at the subgrade layer (l)*
M_G = *maximum volume of temporary stored at the subgrade layer (l)*

$$M_G = M_{GU} \cdot A_H \tag{21}$$

M_{GU} = *maximum volume of temporary stored at the subgrade layer per unit area of the permeable pavement (mm)*

$R_H(t)$ = *volume of horizontal infiltration from the subgrade layer to the infiltration trench (l/Δt)*

f = *horizontal infiltration ratio*

$S_G(t)$ = *infiltration velocity at the subgrade layer (mm/Δt)*

$F_G(_t)$ = *volume of infiltration water from the subgrade layer (l/Δt)*

Infiltration Trench Model

Infiltration

$$V_T' = V_T(t-1) + R_H(t) \tag{22}$$

$$F_T(t) = V_T' \cdot f_I \tag{23}$$

b.   *Flow Out*

$$V_T'' = V_T' - F_T(t) \tag{24}$$

$$Q_T(t) = 0 \qquad (V_T'' < M_T) \tag{25}$$

$$Q_T(t) = (V_T'' - M_T) \cdot f_R \quad (V_T'' \geq M_T) \tag{26}$$

Temporarily Stored

$$V_T(t) = V_T'' - Q_T(t) \tag{27}$$

where,

$F_T(t)$ $=$ *volume of water infiltrated from the infiltration trench (l/Δt)*

$Q_T(t)$ $=$ *volume of water flowed out from the infiltration trench (l/Δt)*

$V_T(t)$ $=$ *volume of temporarily stored at the gravel section of the*
infiltration trench (l)

M_T $=$ *maximum volume temporarily stored at the gravel section of the*
infiltration trench (l)

$$M_T = M_{TU} \cdot A_T \tag{28}$$

A_T $=$ *area of infiltration trench (m^2)*

M_{TU} $=$ *maximum volume of temporarily stored at the gravel section of*
the infiltration trench per unit area (mm)

f_I $=$ *infiltration coefficient*

f_R $=$ *runoff coefficient*

22.4.3 Results of Simulation

Figure 13 shows an example of the comparison between the simulated and the observed surface runoff from the permeable pavement flowing out from the infiltration trench. It shows that the simulated values closely corresponded to the measured values.

The ranges of parameters as follows:

E $= 0.2$ mm/h

S_H $= 15$–35 mm/h (subgrade layer is not saturated)

        $= 25$–45 mm/h (subgrade layer is saturated)

M_{HU} $= 15$–20 mm

S_G $= 1$–5 mm/h

M_{GU} $= 20$–40 mm

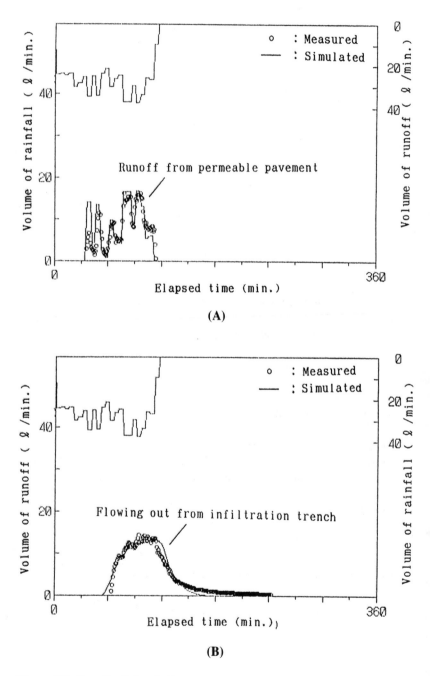

(A)

(B)

Figure 13. Result of simulation.

f $= 0.2–0.5$

M_{TU} $= 1–5$ mm

f_1 $= 0.025–0.050$

f_R $= 0.01–0.02$

Therefore, it is clear that our simulation model can simulate changes of the stormwater surface runoff from the permeable pavement and the stormwater discharge from the infiltration trench. Therefore, with the simulation model we can estimate the effects of the combined infiltration facilities for the stormwater runoff control by infiltration and temporary storage. In particular, the simulation model can represent the quick change of the stormwater surface runoff corresponding with the change of rainfall intensity at the permeable pavement.

The stormwater, which once infiltrated through an infiltration facility, also infiltrates to the other infiltration facility at the combined infiltration facilities. For this reason, we must appraise the effects of the combined infiltration faciltities for stormwater runoff control with due regard to multistage infiltration and runoff processes.

22.5 CONCLUSIONS

The effect and infiltration mechanism of the combined infiltration facilities have been examined. We must appraise the effects of the combined infiltration facilities for stormwater runoff control with due regard to multistage infiltration and runoff processes. For this reason, we have built the multistage simulation model. The conclusions can be summarized as follows:

1. Where the depth of groundwater level is within 1.5 m from the bottom of the infiltration facilities, the initial infiltration heights are considerably influenced by the groundwater level regardless of the type of combined infiltration facility and they have a simple proportional relationship.
2. The final infiltration capacities of the combined infiltration facilities are varied by the infiltration capacities of the ground.
3. The combined infiltration facility of Type 3 exhibits the highest infiltration capacity under the same ground conditions.
4. We can estimate the effects of the combined infiltration facilities for the stormwater runoff control by infiltration and temporary storage with the multistage simulation model.

23 U.S. EPA'S MANUAL OF PRACTICE FOR THE INVESTIGATION OF CROSS-CONNECTION POLLUTION INTO STORM DRAINAGE SYSTEMS

23.1 INTRODUCTION

23.1.1 Objective

This paper summarizes the initial phases of a current U.S. EPA Risk Reduction Engineering Laboratory sponsored research project whose long range goal is to develop a Manual of Practice that will assist local governments to identify the magnitude and source(s) of nonstormwater entries into their separate storm drainage systems.

An important objective of this project is the determination of the most promising techniques to identify, quantify, and locate nonstormwater entries of sanitary and industrial wastes into storm drains. This includes the selection of a mapping and field-screening strategy for the identification of nonstormwater entries, and the evaluation and selection of methods of analyses and investigative parameters for characterizing and tracing contaminating source flows.

23.1.2 Background

Discharges from stormwater outfalls can be a combination of dry-weather base flows; stormwater and snowmelt runoff; intermittent discharges of debris, wash waters, and other waste materials into separate storm drains; and the relatively continuous discharges of sanitary and industrial cross-connected wastewaters. Urban stormwater runoff itself contains the washoff of pollutants from all land surfaces during rains, including storage areas for industrial materials and wastes, gasoline station service areas, parking lots, and other industrial and commercial areas. Consequently, the quality of urban stormwater runoff can vary greatly with climate (dry- vs. wet-weather), time (morning vs. mid-day), season (cold- vs. warm-weather), and location.

As storm drain discharges are often significantly polluted by stormwater alone, the illicit and/or inappropriate discharge of additional wastes into the

0-87371-805-4/93/$0.00+$.50

storm drains can lead to serious water pollution problems. The addition of sanitary wastewaters increases both the concentrations of organic (oxygen demanding) solids and nutrients and the densities of pathogenic microorganisms in the stormwater outfall discharges. Similarly, the discharge of industrial wastewaters and wastes into storm drains can substantially increase the concentrations of many toxic pollutants, e.g., heavy metals and organics. In general, many studies have shown annual discharge loadings from stormwater outfalls to be greatly affected by dry-weather entries.[1-3]

23.1.3   Problem Exemplifications of Nonstormwater Entries into Storm Drainage Systems

Several studies have compared the quality of wet- and dry-weather storm drainage discharges. A Castro Valley, California study found that the urban stormwater runoff was quite soft and had substantial quantities of nonfilterable metals, while the dry-weather flows were very hard and contained few nonfilterable pollutants.[1] However, despite the generally low concentration of pollutants and the low flow rates usually observed in dry-weather flows, the long duration of these flows (e.g., 95% of the time, in many areas of North America) results in the discharge of substantial quantities of many pollutants.

Dry-weather flows in a Toronto, Canada, mixed residential/commercial area were found to have high pesticide concentrations, while a monitored industrial area had dry-weather flows exhibiting high concentrations of organic and metallic toxicants.[3] This study found that more than 50% of the total annual discharges of water volume, total residue, chlorides, and bacteria from the monitored industrial, residential, and commercial areas were associated with dry-weather discharges. Substantial differences in the quality of warm- and cold-weather runoff were also noted.

Another Toronto study of dry-weather flows in separate stormwater outfall discharges in the Humber River watershed attempted to identify the most significant urban stormwater runoff pollutant sources.[4] Approximately 625 outfalls were sampled twice during dry weather and analyzed for several pollutants, including organics, solids, nutrients, metals, phenols, and bacteria. Dry-weather flows were found to contribute significant loadings of nutrients, phenols, and metals compared to upstream conditions. About 10% of the outfalls were considered significant pollutant sources with one third discharging at flow rates greater than 23,000 gal/d (1 L/sec).

Studies have also identified many industrial and sanitary wastewater entries into storm drains. In the Humber River watershed study,[4] an apartment building with sanitary drains from eight units illegally connected to the storm drains was typical of the problems identified. Other problems were found in industrial areas, including liquid dripping from animal hides stored in tannery yards and washdowns of storage yards at meat packing plants. A Bellevue, Washington, study summarized reported incidents of intermittent spills and dumpings of pollutants into the local storm drains.[5] During a 3-year time period, about 50 citizen complaints were made to the Bellevue Storm and Surface Water Utility District regarding water quality problems. About 25% of the complaints concerned oil being discarded into storm drain inlets, while other complaints concerned aesthetic problems, such as turbid or colored water in the creeks. Various industrial and commercial discharges into the storm drains were detected. For example, concrete wastes flushed from concrete transit-mixing trucks and cleaning establishment discharges into creeks were common problems. Vehicle accidents also resulted in spills of gasoline, diesel fuel, hydraulic fluids, and lawn care chemicals that commonly flowed into the storm drain inlets. Relatively significant pollutant yields were also associated with dry-weather discharges from stormwater outfalls in residential areas.[2]

23.2 POTENTIAL DRY-WEATHER DISCHARGE SOURCES

As noted earlier, nonstormwater entries can take multiple forms. Many of these entries are intermittent, occurring during either wet or dry weather. However, as a drainage area increases in size, the probability also increases that dry-weather entries associated with individual intermittent activities would appear more continuous at the outfall. The previously referenced studies commonly found flows at the outfalls during most of the dry-weather period.[1-3]

Examples of intermittent entries include storm runoff from industrial and commercial storage areas, and illegal dumping or washing operations. Continuous nonstormwater entries arise from the connection of industrial, commercial, or sanitary wastewater sources to storm drainage systems. These connections can be associated with "noncontact" cooling waters (which frequently contain a variety of chemical substances, including algicides and corrosion inhibitors), other industrial wastewaters, and sanitary wastewaters. The following list summarizes the potential contaminated residential area nonstormwater sources being evaluated. Commercial and industrial dry-weather discharges are being addressed in a separate study that will be incorporated into the proposed EPA Manual of Practice.[6]

Sanitary wastewater:
- raw wastewater from directly connected or leaky sanitary sewerage
- effluent from improperly operating septic tank systems

Household automobile maintenance:
- car washing
- radiator flushing
- engine degreasing
- improper oil disposal

Lawn irrigation:
- over-watering residential areas, parklands, and golf courses
- direct spraying onto impervious surfaces

Roadway and other accidents:
- fuel spills
- truck spills
- pipeline leaks and releases

Other:
- washing of concrete transit-mixing trucks
- laundry wastewaters
- improper disposal of household toxic substances
- dewatering of construction sites
- sump-pump discharges
- contaminated surface waters and groundwaters
- automobile service station drippings and leaks
- underground storage tank leaks

Outfall discharges can be separated and ranked into three general categories which can enable the identification of those outfalls requiring further analyses and investigations. These categories are (1) pathogenic or toxic pollutant sources, (2) nuisance and aquatic life threatening pollutant sources, and (3) unpolluted water sources.

The pathogenic and toxic pollutant source category is considered the most severe and could cause disease upon water contact or consumption and significant impacts on receiving-water biota. These sources may also cause significant water treatment problems for downstream consumers, particularly when the pollutants include soluble metal and organic toxicants originating from sanitary, commercial, or industrial nonstormwater discharges. Other residential area toxic pollutant sources include inappropriate household toxicant disposal, automobile engine degreasing, vehicle accident cleanup, and irrigation runoff from landscaped areas excessively treated with fertilizers and/or pesticides.

Nuisance and aquatic life threatening pollutant sources from residential areas include laundry wastes, landscaped areas irrigation runoff, automobile washing, construction site dewatering, and washing of concrete transit-mixing trucks. These pollutants can cause excessive algal growths; tastes and odors in water supplies; and highly colored, turbid, or odorous waters.

Clean water discharged from stormwater outfalls can originate from natural springs feeding urban creeks that have been converted to storm drains, infiltrating groundwater, and infiltrating domestic water from waterline leaks.

23.3 NONSTORMWATER ENTRY IDENTIFICATION STRATEGY

23.3.1 Initial Mapping Effort

The preparation and study of drainage and land-use maps are an important step in a nonstormwater contamination investigation; therefore, before field activities can be started, a mapping effort must be undertaken. The first objective of the mapping effort is the identification and location of all stormwater outfalls. Locating outfalls is not a trivial procedure. In previous case studies, repeated field trips typically uncovered additional outfalls that were not located during earlier trips. In Toronto, for example, while most outfalls were located during the first field trip, two more trips were needed to locate all outfalls.[4] Since communities do not often maintain up-to-date mapping of drainage facilities, it is very common to find outfalls in the field that do not appear on storm drainage system maps.

Another important objective of the initial mapping effort is the identification of the drainage areas tributary to each outfall. Drainage maps should identify: (1) predevelopment streams that may have been converted to storm drains as well as, (2) all current and past land uses. Specific land-use categories that must get special attention include commercial and industrial plus others that may contribute runoff problems, e.g., landfills. Any industrial activities having significant potential of contributing flows to the storm drainage system must be identified and located.[6]

Further drainage area investigations must be conducted if outfall screenings indicate the presence of dry-weather discharge. These include drainage system and industrial and commercial site studies (such as flow, dye, and smoke studies) to locate specific nonstormwater entries.

Aerial photography can be a useful tool for nonstormwater entry detection. For example, aerial photography can be used to identify residential areas containing failing septic systems; continuous discharges to surface drainages, such as sump discharges; and storage areas that may be contributing significant amounts of pollutants to stormwater runoff.

23.3.2 Initial Field Surveys

After all background data has been collected, initial field screening surveys can be undertaken to identify those outfalls needing more detailed investigations for pollutant source identification and control options. Field surveys include visual inspection of the outfall and its vicinity and physical and limited chemical evaluations of any outfall discharges. In addition, as many of the dry-weather discharges are intermittent and may not be detected during any single investigation, it is important to survey during anticipated peak sanitary wastewater discharges and to resurvey all outfalls over a period of time including several seasons. Repeated surveys will significantly increase the probability of identifying outfalls containing dry-weather entries. Various physical characteristics near the outfall can also provide evidence that inappropriate discharges periodically occur, e.g., the presence of stains, debris, structural damage or corrosion, or unusual plant growth or the absence of plants. Intermittent flows and/or debris can be trapped between outfall visits using: (1) small caulk dams placed in the storm sewer inverts, or (2) coarse screens, respectively. Alternatively, automatic samplers, operating on a 15-min basis for example, can operate over a 24-h period to detect and sample intermittent flows at suspected outfalls.

Initial field-screening activities currently being evaluated include:

- placement of outfall identification number
- rough estimate of outfall discharge
- inspection of the outfall area for "tell-tale" signs, e.g., the observance of floatables, coarse solids, color, oil sheens, and odor characteristics of discharge and/or receiving water; dead fish or aquatic vegetation in the receiving water; and stains, debris, damage to concrete, corrosion, unusual vegetation, etc. of the storm sewer, outfall facility, or receiving-stream embankments and channel
- measurement of water temperature
- collection of water samples for laboratory analyses

23.4 EVALUATION OF NONSTORMWATER CONTAMINATION ANALYSES

An important goal of this project is the evaluation and comparison of standard and alternative analytical methods and the selection of parameters (tracers) for identifying contaminating source flows. Alternative laboratory and field analyses were identified and tested for each tracer during the first phase of this research. In general, the selection of an analytical method to determine a tracer concentration will depend upon many conditions, most notably the expected tracer concentration distribution (mean and variance) in both the uncontaminated base flow and the potential nonstormwater source flows. Other factors affecting procedure selection include detection limit and repeatability, ease-of-use, analyses interferences, cost of equipment, training requirements, analyses time requirements, and the required minimum contamination level to be identified. The potential tracer parameters and the sampling and laboratory procedures used to evaluate them are summarized below.

23.4.1 Dry-Weather Discharge Parameters

Where dry-weather-flow discharges were observed during the field evaluation, and water samples were collected for later analyses. Laboratory analyses included the measurement of:

- water color
- pH, conductivity, and hardness concentration
- chlorine and fluoride concentrations
- ammonia and potassium concentrations
- surfactant (detergent) concentration and fluorescence
- specific toxicants (priority pollutant) (arsenic, copper, chromium, cyanide,

phenols, zinc, PAHs, phenols, and phthalate esters) concentrations and toxicity screening (using a Microtox® procedure)
• molecular tracers (coprostanol and epicoprostanol)

Many of the parameters listed above were chosen for further evaluation on the basis of earlier case study results and characterization information for potential source flows. Other parameters are included in the EPA's Final Rule, National Pollutant Discharge Elimination System Permit Application Regulations for Storm Water Discharges.[7] Notably absent from the above list are bacteria and dissolved oxygen, which have been shown in previous nonstormwater entry studies to provide limited useful information.

23.4 Sampling and Laboratory Methods Investigation

Initially dry-weather flows were sampled at 12 locations from a grass swale drainage system serving a residential area containing septic tanks. Samples were obtained during an excessively dry summer period. Each of these 12 samples were subjected to approximately 35 different tests which compared several analytical methods for each of the major tracer parameters of interest. Tests were conducted to enable comparison of the results of alternative tests with standard procedures and to identify which methods had suitable detection limits, based on real samples. In addition, four representative samples from this grass swale/septic tank residential area were further examined using standard addition methods (known amounts of standards added to the sample and results compared to unaltered samples) in order to identify matrix interferences. Matrix interferences are generally caused by contaminants in the samples interfering with the analysis of interest. Many of the analysis methods were also tested against a series of standard solutions to identify linearity and detection limits. After the most suitable analytical methods were identified, known mixtures of potential contaminating flow sources (septage, sanitary wastewater, cooling water, and treated and untreated plating-bath waters) with local spring waters were prepared and analyzed. Mixtures contained these contaminated waters mixed with spring water in the following percentages: 0, 0.1, 1, 5, 10, 25, 50, 75, 90, 95, 99, 99.9, and 100. These known mixtures were analyzed using the selected chemical methods to determine the errors and useful detection limits associated with each analytical method.

Many source flow samples (sanitary wastewater, septage, carwash water, laundry waters, construction site dewatering waters, metal plating-bath liquids, cooling waters, etc.) are currently being collected and analyzed to identify typical variations in tracer concentrations. The tracer concentration variations will have a significant effect on the ability to statistically identify small contributions of contaminated source flows in clean base flows. Statistical analyses were conducted during previous research phases using a Monte Carlo model and limited data on tracer concentration variability. These analyses showed that source flow tracer concentration variations and analytical detection limits significantly affect the ability to detect small contamination levels that may cause serious receiving water problems. Large variations in tracer concentration (not explained by examining seasonal trends or other factors), can render the identification of contaminated sources difficult, especially if these sources are diluted by other flows.

Sixty-six outfalls in a mixed land-use area are currently being examined over one year to test selected parameters for use in identifying contaminating sources. In addition, selected outfalls are being frequently examined using automatic sampling equipment to determine the presence or absence of outfall flow and variations in quality. Knowledge of typical variations in outfall conditions will enable the determination of required outfall screening frequency to satisfactorily detect low levels of storm drainage contamination during dry weather.

23.5 RESULTS AND RECOMMENDATIONS

23.5.1 Screening Activities

Past and present research indicates that basic field activities such as simple outfall flow rate estimates, noting the presence of oil sheens, coarse solids, floatables, color, odors, etc., are very worthwhile in identifying problem outfalls that require further investigation. As emphasized earlier, outfalls exhibiting the signs of noncontinuous discharges should be visited several times to increase the probability of observing and sampling a dry-weather discharge. Analyzing pooled water immediately below the outfall or collected between visits in small, constructed dams within the storm sewer can greatly assist in identifying noncontinuous discharges. Similarly, coarse solids and/or floatables can be similarly captured through the erection of coarse screens and/or booms at the mouth of the outfall or in the receiving stream.

23.5.2 Sampling Strategy

The importance of sampling all outfalls, regardless of size, was established during the early phases of this study. Presently, 66 outfalls in a residential and commercial area are being evaluated in detail. Of this group the median outfall size is 16 in., and more than 75% of the outfalls are less than 36 in. in diameter. About one quarter (17) of these 66 outfalls are consistently flowing, with about two thirds of the flows discharging from pipes that are less than 36 in. in diameter. Of the 66 outfalls, four exhibit dry-weather flows which are extremely toxic or are raw undiluted sanitary wastewaters. Each of these four contaminated outfalls is 20 in. or less in diameter. Some of the worst dry-weather-flow discharge problems were associated with very small (4-in. diameter) pipes draining automobile service areas. This demonstrates that smaller outfalls can contribute significant pollutant loads to receiving waters and should not be neglected if receiving water improvement is a serious goal of a cross-connection investigation.

Examination of the 66 outfalls during the first three separate sampling occasions found that while some of the dry-weather flows occur intermittently, most are continuous. Automatic samplers were also being used at several of the outfall sites to determine the visiting/sampling frequency needed to identify all problem outfalls with a reasonable degree of certainty.

23.5.3 Nonstormwater Contamination Analyses

Selected measurements are recommended for a minimum and routine outfall investigation. Determination of the following parameters, utilizing appropriate procedures, will generally result in adequate information for the detection and identification of major pollutant sources:

- water color
- pH, conductivity, and hardness concentration
- fluoride concentration
- ammonia and potassium concentrations
- surfactant (detergent) concentration and fluorescence
- toxicity screening

Parameters which have been removed from the initial list include chlorine concentrations, specific toxicant (arsenic, copper, chromium, cyanide, phenols, zinc, PAHs, phenols, and phthalate esters) concentrations, and molecular tracer concentrations. These parameters where proven infeasible, even if correctly measured, since they fail to provide continuously useful information for the identification of inappropriate storm-drainage-system entries. In descending order of importance, the primary reasons for rejection of analytical procedures or parameters were: insufficient detection limits; inconsistent results; matrix interferences; highly variable concentrations in source flows; presence in multiple source flows requiring separation; and difficulty and/or expense involved.

23.5.4 Selected Analytical Methods and Procedures

Most of the recommended analyses are conducted using small "field-type" instruments. However, despite their portability, the use of these instruments in the field can introduce many errors. Currently, temperature and conductivity are the only analyses that are routinely conducted in the field. For other analyses, samples are collected at the site, iced, and taken back to the laboratory for study. At each outfall an approximate 0.5-gal (2-L) sample of dry-weather discharge is collected and stored in a plastic container. Additionally, a second 250-mL sample is collected in a Teflon™-lidded glass bottle for organic analyses and the toxicity screening. All samples are analyzed (or extracted) within accepted time limits.

The following describes the procedures and parameters that are currently being used for the analysis and identification of dry-weather discharges:

- Water color is determined in the laboratory using a simple comparative colorimetric field test kit.
- pH, which can be influenced by various industrial discharges, is measured in the laboratory using a standard pH meter. pH measurements using pH test paper have been found to be within one unit of the laboratory meter. However, this difference is believed to be too large and an accurately calibrated pH meter used in the laboratory on fresh samples is the recommended test procedure. Small "pen" pH meters most suitable for field use can easily be inaccurate by a 0.5 pH unit and are relatively hard to calibrate, and accordingly must be used with care.
- Conductivity and temperature are quickly and easily measured using a dual dedicated meter in the field.
- Fluorides are easily detected in the laboratory using a field spectrometer and evacuated reagent and sample vessels (HACH DR/2000 and AccuVac). The AccuVac procedure worked well for sample concentrations <2.5 mg/L; however, in the rare instance of higher concentrations, sample dilution is required because of nonlinear responses. Specific-ion probes were also evaluated, but the technique proved to be too inconsistent, especially for personnel having little training.
- Ammonia can be easily measured in the laboratory using a direct Nesslerization procedure and a field spectrophotometer. The use of various indicator test papers and simple field test kits for ammonia determination gave poor results. Specific-ion probes were also tested. Typical problems encountered were associated with color interferences, long analysis times, inconsistent results, and poor performance when standard solutions were analyzed.
- Potassium is measured in the laboratory either using a field spectrophotometer or flame atomic absorption. Specific-ion probes were also evaluated and indicated the same poor results found for fluorides and ammonia.
- Detergents are measured in the laboratory using a field comparative colorimetric method with a detection limit of 50 µg/L. Fluorescence is also being analyzed using a fluorometer. Specific-ion probe titrations for detergents were not successful because of poor detection limits.
- Hardness is determined in the laboratory using a field-titrimetric kit. A number of simple field test kits were tested but the direct reading titration method proved most convenient and accurate. However, hardness test paper is being used to estimate the titration end point.
- Toxicity-screening tests have been found to be very useful as indicators of inappropriate nonstormwater entries. The Microtox® toxicity screening test

is the method being used. If a sample results in a large toxic response, then specific toxicant analyses (organics and metals) could be performed to better identify the toxicant source. A number of simple test kits were used for specific heavy metal analyses, but with very poor results. High-detection limits and interferences make these methods impractical, unless an outfall is grossly contaminated with a concentrated source such as raw plating bath wastewater. Another recommended method for toxicant identification of metals would be an atomic emission spectrophotometer, or a similar instrument which can routinely analyze a large number of metals. A base/neutrals/ acid (BNA) scan using a gas chromatograph with a mass spectrophotometer detector is also being routinely conducted on all samples. In general, the Microtox® screening test was found to be an efficient method of toxicity analysis, particularly for identifying samples requiring further analyses.

Data Evaluation

Pollutant contributions associated with nonstormwater entries can often be distinguished through the unique characteristics of various water types and typical associated tracer concentrations. The analysis scheme recommended above should allow an efficient determination of the general category (toxic/ pathogenic, nuisance, or clean) of the water being discharged. For example, major ions or other chemical/physical characteristics of the flow components can vary substantially depending upon whether the water supply source(s) are groundwater or surface water, and whether the source(s) are treated or not. Fluoride can often be used to separate treated water from untreated water sources. This latter group may include local springs, groundwater, regional surface flows, or nonpotable industrial waters. If the treated water has no fluoride added, or if the natural water has fluoride concentrations close to potable water fluoride concentrations, then fluoride may not be an appropriate indicator. Hardness can also be used as an indicator if the uncontaminated base flow is hard groundwater, and the treated water source is from soft surface supplies.

Water from potable water supplies (that test positive for fluorides or other suitable tracers) can be relatively uncontaminated, e.g., domestic waterline leakage or irrigation runoff, or heavily contaminated, e.g., sanitary wastewater.

In areas containing no industrial or commercial sources, sanitary wastewater is probably the most important nonstormwater source. Surfactant measurements may be useful in determining the presence of sanitary wastewaters. However, surfactants present in water from potable water sources could indicate sanitary wastewaters, laundry wastewaters, car washing water, or any other waters containing detergents. The presence of fabric whiteners (as measured by fluorescence using a fluorometer in the laboratory or in the field) can also be used in distinguishing laundry and sanitary wastewaters. If surfactants are not present, then the potable water could be relatively uncontaminated, e.g., domestic waterline leaks or irrigation runoff.

Sanitary wastewaters often exhibit relatively consistent characteristics, e.g., volume and constituent concentrations. The ratio of ammonia to potassium has been found to be an effective indicator of sanitary vs. septic tank wastewaters.[8] If the surfactant concentrations are high, but the ammonia and potassium concentrations are low, then the contaminated source is likely laundry wastewater. Conversely, if ammonia, potassium, and surfactant concentrations are all high, then sanitary wastewater is the likely source. Obviously, odor and other physical appearances such as turbidity, coarse and floating "telltale" solids, foaming, color, and temperature would also be very useful in distinguishing sanitary wastewater from rinse water or laundry wastewater sources.

23.5.6 Confirmatory Analyses

Several confirmatory analyses are being studied for usefulness in verifying the more significant sources of nonstormwater discharges. These analyses can be very useful to check for false negatives and obtain more specific results on a random basis. However, these analyses require highly trained personnel and specialized equipment that would generally not be available in most laboratories. Consequently, it may not be feasible to perform confirmatory analyses on the large number of samples collected from several hundred outfalls several times a year. The confirmatory analyses currently being evaluated include:

- trihalomethanes
- specific bacteria
- molecular markers

Trihalomethanes (THMs) are formed when chlorine reacts with certain natural organics present in water. The detection of these compounds in groundwaters has been used as a positive indication of chlorinated potable-water leakage.[9] Chloroform and dichlorobromethane are the THMs most frequently monitored because of their low detection limits and their specific indication of potable water.

Bacteria are generally unreliable indicators of wastewater sources. Past use of fecal streptococcus to fecal coliform ratios to distinguish human from nonhuman bacteria sources has not proven very successful, especially for mixed or aged wastewaters. Other specific bacteria types, e.g., streptococcus biotypes may be useful as potential indicators of human sewage. Another indicator of human wastes may be the use of certain human-specific molecular markers, specifically the linear alkylbenzenes.[10] Fecal sterols, e.g., coprostanol and epicoprostanol have been used for the detection of raw sewage in marine waters;[10] however, evaluations of their usefulness conducted during this research in identifying nonstormwater contamination have not shown them to have the promise originally thought.

23.5.7 Source Identification

After initial outfall surveys have indicated the presence of contamination, further detailed analyses are needed to identify and locate the specific contaminant source(s) (e.g., residential, commercial, and/or industrial) in the drainage area. For source identification and location, upstream survey techniques should be used in conjunction with an in-depth watershed evaluation. Information on watershed activities can be obtained from aerial photography and/or zoning maps, while upstream survey techniques will include:

- the analysis of the dry-weather flow at several manhole points along the storm drainage system to narrow the location of the contaminating source;
- tests for specific pollutants or ions associated with known activities within the outfall catchment area; and
- the measurement of water flowrate and temperature, visual and television inspections, and smoke and dye tests.

For the analysis and identification of potential industrial discharge sources, a description of these and other identification methods is contained in a separate study which specifically addresses commercial and industrial nonstormwater entries into storm drainage.[6] In addition, future work will address the source identification and location procedure in detail.

23.6 CONCLUSION

Many urban storm runoff studies have found that dry-weather discharges from stormwater outfalls can contribute significant pollutant loadings. Ignoring these loadings can lead to improper conclusions concerning stormwater control requirements.

Municipalities that have recognized the importance of dry-weather flows have attempted to investigate their sources using various methods. Given the very large numbers of outfalls in most municipalities, a prerequisite of any method is an efficient means by which to distinguish those outfalls creating the most severe problems. This project seeks to characterize and rank outfalls into one of three categories: pathogenic/toxicant, nuisance and aquatic life-threatening, or clean water. The first, and most important category, generally describes dry-weather contaminated flows whose sources are sanitary or industrial wastewaters. The initial screening of all outfalls and the analysis of all dry-weather flows provide a high probability of identifying outfalls in this most critical category.

Ongoing and future research will seek to evaluate and refine the recommended field and laboratory screening procedures through test surveys in a number of diverse watersheds. Future project phases will also address procedures to correct specific nonstormwater discharges once they are identified.

ACKNOWLEDGMENTS

This project is being carried out by the Civil Engineering Department of the University of Alabama at Birmingham under the sponsorship of the EPA. The opinions expressed in this paper are the opinions of the authors alone, and are not official policies of the EPA.

REFERENCES

1. Pitt, R., and G. Shawley. "A demonstration of nonpoint pollution management on castro valley creek," U.S. EPA Report-NTIS No. PB 83-163-089 (1982).
2. Pitt, R. "Characterizing and controlling urban runoff through street and sewerage cleaning," U.S. EPA Report-/600/2-85/038, NTIS No. PB 85-186-500 (1985).
3. Pitt, R., and J. McLean. "Humber River pilot watershed project," prepared for the Ontario Ministry of the Environment, Toronto, Ontario (1986).
4. Gartner Lee and Associates, Ltd. "Toronto area watershed management strategy study, technical report #1, Humber River and Tributary dry-weather outfall study," Ontario Ministry of the Environment, Toronto, Ontario (1983).
5. Pitt, R., and P. Bissonnette. "Bellevue urban runoff program, Summary Report," U.S. Environmental Protection Agency and the Storm and Surface Water Utility, Bellevue, Washington (1984).
6. Cadmus Group, Inc., and Triad Engineering, Inc. "Industrial non-stormwater discharges," U.S. EPA Report, Contract No. 68-c9-0033 (1990).
7. "National pollutant discharge elimination system permit application regulations for storm water discharges," U.S. EPA Final Rule, 40 CFR Parts 122-124 (November 16, 1990).
8. Pitt, R., M. Lalor, M. Miller and G. Driscoll. "Assessment of nonstormwater discharges into separate storm drainage systems: protocol for the proposed manual of practice," U.S. EPA Draft Report, Contract No. 68-C9-0033 (1990).
9. Hargesheimer, E.E. "Identifying water main leaks with trihalomethane tracers," *J. Amer. Water Works Assoc.*, 67:71 (1975).
10. Eganhouse, R.P., D.P. Olaguer, B.R. Gould and C.S. Phinney. "Use of molecular markers for the detection of municipal sewage sludge at sea," *Mar. Environ. Res.* 25:1 (1988).

INDEX

A

Action plans, in catchment management, 72–75
Activated carbon adsorption, in urban stormwater reclamation, 324–325
Aesthetic deterioration, receiving water and, 7
Agriculture, pollution transport to coastal waters and, 90
Annual statistics, *see* Yearly entries
Aquaculture, urban stormwater reclamation for, 299–305
  materials and methods in, 300–301
  results of, 301–305
Aquatic environment, 60
Asphalt-concrete lot, conventional, 154, 158–159
Asphalt lot, porous, 152–153, 156
Attenuation storage, source control and, 187–188

B

Bacteria, *see also* Pathogens
  coliform, receiving water and, 7, 9
  numeric criteria for development of, 226–229
  in Massachusetts, 214
Baltic Sea, 87
Base flows, reduction of, urbanization and, 260–261
Baseline surveys, catchment management and, 71–72
Best Available Technology/Uniform Emission Standard, 60
Best management practices

level spreader system and, 102
for urban stormwater runoff control, 191–203
  detention ponds, 191–196
  infiltration facilities, 196–198
  mining ponds, 202–203
  vegetative filter strips, 198–201
  wetlands, 201–202
Biological impacts, receiving water and, 9
Biological treatment, 30
  in urban stormwater reclamation, 321, 324
Boat moorings, 73
Bukit Timah flood alleviation scheme, hydrologic analysis for, 279–287
  alternatives evaluated with, 286–287
  models for, 281–284
  objectives of, 280–281
  recommendations based on, 287
  system evaluation with, 284–286

C

Carbon adsorption, in urban stormwater reclamation, 324–325
Catchbasins, 20
Catchment issues, 71
Catchment management planning, 62–64
  Isle of Man, *see* Isle of Man communities, pollution control management plan for
  U.K. experience with, 67–77
    action plans in, 72–75
    baseline surveys and, 71–72
    issues in, 71
    National Rivers Authority and,